Abstract. Due to some subjective and objective circumstances, the author was directly involved as a Researcher in experiments with mitochondria for more than 50 years. This book is an attempt to share with the reader the unique experiences in resolving problems, which always arise during experimental work. The book describes the methods of isolation of mitochondria from different organs of normal and diseased animals and the methods to study the major mitochondrial functions: respiration, membrane potential, ion transport, calcium-dependent permeability transition, production of reactive oxygen species. Some problems of the contemporary research on mitochondria are technical, such as the problems with the quality of water and chemicals. However, the major problems are caused by outdated methodology and some of the methods to study mitochondrial functions. Therefore, in addition to description of technical details, the author also gives, although only briefly, the theoretical backgrounds underlying the methods and functions. The author believes that mitochondria have to be studied in close relation to the functions of related organs. After all, mitochondria are not "just another subcellular organelles", Mitochondria are "The Fundamental Organelles", which determine Life and Death of most living organisms on the Earth.

I0471301

FOREWORD

I wrote this "Manual" keeping in mind the logic of Methods rather than following the logic of Mitochondria. Therefore, description and discussion of mitochondrial functions are related to the Methods and Methodology under consideration. My intention was not just to share with young researchers my encounters with problems in Experimental Mitochondriology, which I experienced during my 50 years "carrier" as a scientist, but also to share related stories, thoughts and ideas, which also are bound to Methodology.

Over the years, I have accumulated many protocols, methods, references, and "know how". It would be a shame if this experience and knowledge was lost. So, I decided to present and explain these protocols to young researchers, and save them time by explaining pitfalls of methods and warn against many "conventional methods". It is by doing experiments that you can be at the edge of unknown and finding new facets of the Nature. Therefore, doing experiments is the real Challenge and Fun of being a Scientist. Biochemistry, above all, is experimental science. Therefore, when you do experiments you are constantly challenged with a lot of problems that must be resolved in order to make the results of your experiments credible.

By acquiring experience and solving problems you begin critically regard publications of other researchers, unlike too many "Bosses", who believe everything what is published. In fact, you will be astonished by realization how much crap is published and the amount of money that is wasted on results obtained with "conventional" but often obsolete methods, and on "Big" ideas that lead nowhere (I mean the so called "-omics"). You have to educate yourself on the problem you are going to work on. For this, you have to read a lot of papers. First, pay attention to the section describing Methods. You will save a lot of time not reading papers, which contain methodical errors. For example, there are so many papers describing the roles of mitochondria in various pathologies with the results obtained on cultured cells that were grown in the presence of antibiotics penicillin and streptomycin. It is known that aminoglycoside antibiotics (gentamicin, streptomycin) make mitochondria incapable of respiration.

At some points, when I looked back on the results obtained earlier by so many respected researchers, I realized that many results and conclusions made, say 10 or 20 years ago, were wrong. The simple fact is that when you work at the Edge of the Known, you make mistakes because you enter the Realm of Unknown. Therefore everybody makes mistakes, but you have to avoid those errors, which result from bad planning and narrow-mindedness or ignorance. Some of the reasons that led to erroneous conclusions in the past will be discussed in this Manual.

In each Chapter the figures and tables are numbered consecutively, for example, Fig. 6.1 and Table 6.2 are figure 1 and Table 2 in Chapter 6.

I also ask the readers for responses. I will appreciate if you will show me any errors, unclear parts, or write me if you have questions and suggestions. I hope that this little book will be helpful to you in your experimental work. Have Fun!

My e-mail for communication is: *alexander.panov55@ gmail.com*

Acknowledgements.

My greatest Gratitude is to the late Professor Lars Ernster whom I regard as my Scientific Farther. Lars taught me the importance of Controls. Then, I thank Key LaNoue and Russell Scaduto, who helped me to move from disintegrating USSR to USA. The most productive years in the USA were the years at Emory University in Atlanta, Georgia. I worked under Professor Tim Greenamyer, and studied the roles of mitochondria in pathogenesis of Neurodegenerative diseases. Tim provided everything I needed for my research. Moreover, when Tim moved from Emory and Atlanta, he left as a gift all the equipment and supplies, which I had at the time. Because of that, the next eight years were also productive. Thank you, Tim!

CHAPTER 1

Water and Chemicals

Before you start doing experiments, you have to prepare a large number of different solutions: isolation and incubation buffers, stock solutions of substrates and inhibitors. And while doing this "simple" job, you may encounter many unexpected problems, which will influence the final outcome of your experiments. Most solutions use water as a solvent. Therefore, I begin my story with water.

Water.

The quality of water during the last two decades became one of the most important determinants of the quality of isolated mitochondria. A little more than a couple of decades ago the commonly used water for biological studies was the glass double distilled water. The major drawbacks of water distillers are high energy cost, low productivity and a very large waste of water for cooling of water vapors. Understandably, with diminishing water supply and rising energy cost, new much more productive water purification systems have been developed and introduced into practically all research laboratories over the World; Together with the new problems.

When I first made an attempt to describe the problem with water in one of my papers, the Editor regarded it as "anecdotal". Later, when I worked at the Carolinas Medical Center, I several times raised the problem before the "Big Boss". Finally, he angrily blurted that he is fed up with this "stupid water problem". This is a typical response of people who forgot, or never knew, what the Experimental Work is. Now, at least you will listen to my story about the problems with water. The seriousness of this problem led to the fact that during the last several years you will not find many papers in which researchers would present data on mitochondrial respiration.

The most popular contemporary systems for purification of water use several steps of water filtration, deionization with the cation and anion exchange resins, and in some systems, radiation of water with ultraviolet. Often, this type of water is called "Milli-Q water". "The term is commonly (mis)used as a generacised trademark to refer to other purified waters (and purification equipment), whether from other systems manufactured by Millipore (such as *Elix*), or from systems manufactured by other companies (Siemens, for example). When used in this fashion the term's meaning is less precise" (Cited from the Wikipedia's article on "Milli-Q water").

Milli-Q water can be contaminated with divalent cations other than Ca^{2+}, presumably Cu^{2+} or Zn^{2+}.

According to Wikipedia, Milli-Q is a trademark created by Millipore Corporation to describe 'ultrapure' water of "Type 1", as defined by various authors (e.g. ISO 3696), as well as their devices for producing such water [1]. The purification processes involve successive steps of filtration and deionization, to achieve a purity expediently characterized in terms of resistivity (typically 18.2 MΩ cm). When used in this fashion, the term has a fairly well-defined meaning (Wikipedia).

Milli-Q water uses resin filters and deionization to purify water (usually the tap water). The system monitors the ion concentration by measuring the electrical resistivity of the water. Higher resistivity means fewer charge-carrying ions. Most *Milli-Q* systems dispense the purified water through a 0.22 μm membrane filter.

In 1997 I was working at Emory University at the Center for Molecular Medicine under Douglas C. Wallace, the father of the "Mitochondrial Medicine". At the time, I was studying mitochondria from prostate cancer cells, and as a control and for calibrating the methods I used freshly isolated mouse liver mitochondria. One day, I noticed that my liver mitochondria were deteriorating very rapidly. To make the story short, although it took a lot of work and time, I have found that the Milli-Q water, which I used for preparation of my isolation and incubation buffers, was contaminated with divalent cations, other than Ca^{2+}, because not EGTA but EDTA and citrate eliminated the toxicity from my buffers. Taking into consideration that in the US many pipes in the buildings are made of copper, and that the Milli-Q water had electrical resistance of 18.2 MΩ cm, and the effectiveness of EDTA, I suggested that the Milli-Q system's water was contaminated with Cu^{2+} ions at very low concentration. I began to make my solutions using the glass double distilled water, which was also available in the lab, and the quality of my mitochondria returned back to normal. If I would stick from the beginning to the bidistilled water, I would never have the problem. However, I began to use the Milli-Q water because everyone around was telling that this is the MOST PURE water, it was "cool" to use it instead of "old" BiDi. But I did encounter the problem; half solved it, and thus saved myself from many problems in the future.

In recent years, several of my colleagues complained to me that they had problems with the sudden decrease in the quality of mitochondria. In all cases they used the Milli-Q water. After my colleagues followed my suggestion to use bidistilled water instead, the problems have been resolved. During my working at different institutions I observed that some of the Milli-Q water systems provided water with more or less acceptable quality. But then, suddenly, the water becomes toxic.

Still, I wandered how very tiny amounts of divalent cations might be so toxic and destroy mitochondria just in less than an hour after isolation? It was shown that Cu and Zn ions at concentrations as low as 10^{-12} M were toxic to mitochondria (Brewer 2007). The answer came to me a few years later, when I started to study mitochondrial production of reactive oxygen species.

The Milli-Q-water is always contaminated with hydrogen peroxide.

Brain and heart mitochondria have very high activity of superoxide dismutase 2 (SOD2) located in the matrix space. The enzyme catalyzes dismutation of superoxide radical (O_2^{*-}) to hydrogen peroxide: $O_2^{*} + 2H^+ + e^- \rightarrow H_2O_2$

Dismutation can proceed spontaneously, but SOD accelerates it dramatically. I measured production of superoxide radical by brain mitochondria using the Amplex red method described below. The method is based on the reaction catalyzed by horse radish peroxidase (HRP): Amplex Red + H_2O_2 → Resorufin

Resorufin has a very strong fluorescence. The stoichiometry is 1 resorufin molecule formed per 1 molecule of H_2O_2.

I have measured the fluorescence of resorufin using fluorimeter made by the C&L company, located in Middletown, PA (www.fluorescence.com), established by my first "American Boss" Russell Scaduto. The "C" and "L" in the company's Logo stand for "Cara" and "Lora" – Russell's beautiful daughters. The instrument used two wheels with 8 filters for emission and absorption, and had very high sensitivity – the photomultiplier could register a single photon! Therefore, in order to prevent damages to the photomultiplier at very high fluorescence, the instrument had a cutoff at the fluorescence intensity of 60,000 arbitrary units. Unlike commercial spectrofluorimeters, which zero the background fluorescence before measuring fluorescence of a sample, my instrument measured the sum of fluorescences of a sample and the background. I had to subtract the background fluorescence later during calculations. Usually, the general background fluorescence was around 20,000-35,000 of arbitrary units. However, when I started to work at the Carolinas Medical Center (Charlotte, NC), the Milli-Q water gave fluorescence of 45,000 AU, or higher. The high background fluorescence often did not allowed me to perform experiments because it was too close to the cutoff. So, I studied the problem of the background fluorescence more closely. After trying water from different Milli-Q systems and the very expensive and sophisticated water purifying system at the Hospital of the Carolinas Medical Center, I concluded that the fluorescence originated from the ion exchange resins. The fluorescence of water would also significantly increase after I placed water or my media into a new plastic tube, or sterilize my solutions using the Millipore filters. The fluorescence would decline to a low level (about 700-1000 AU) after I added an aliquot of catalase. This proved that the water fluorescence was due to the presence of hydrogen peroxide. During preparation of the isolation or incubation media (I measured the fluorescence changes after addition of each component) I found that some chemicals would also significantly increase the fluorescence while others did not. Using calibration with resorufin I estimated that even the best Milli-Q systems produced water, which contained from 50 to 250 nM (or even higher) hydrogen peroxide!

This fact explained the toxicity of Milli-Q water for mitochondria if it was contaminated by even a tiny amount of Cu^{2+} or Zn^{2+} ions. In the presence of hydrogen peroxide these transition cations stimulate the Fenton's reaction, which produces highly toxic hydroxyl radical. In this respect Cu^{2+} and Zn^{2+} are 1000 times more active than Fe^{2+} (Dikalov et al., 2004).

Nonenzymatic interaction of a transition metal with hydrogen peroxide produces the highly reactive hydroxyl radical (OH^*). Addition of a reducing agent, such as ascorbate, leads to a cycle which increases the damage to biological molecules.

$$Fe^{2+} + H_2O_2 \longrightarrow Fe^{3+} + OH^* + OH^-$$

The hydroxyl radical interacts with any molecule it encounters, and stimulates lipid peroxidation. Thus, in the presence of high levels of H_2O_2 in the isolation or incubation buffer, the tiny concentrations of Cu or Zn ions cause massive formation of OH^* that rapidly damage mitochondria. I should mention here, that even in the absence of water contamination with transition metals, the presence of 50-250 nM of hydrogen peroxide in the buffers damaged brain and spinal cord mitochondria isolated from the transgenic SOD1 animals.

Solutions to the problems of water contamination with hydrogen peroxide and Cu and Zn ions

The best solution of the problems associated with water quality would be to use the glass bidistilled water. However, from my experience I know that many administrators will regard your problems as anecdotal and will not spend money for expensive distillers. Even to assemble a distiller from separate components would cost in the range of several thousand dollars because the prices for the glass coolers, connectors and Pyrex flasks are too high. Therefore, I found several simple, inexpensive solutions, which allow consistently isolate mitochondria with high respiratory rates and respiratory control ratios above 5-6.

It is also important to store water and all your solutions in glass bottles and containers closed tightly.

Figure 1.1. Home distiller.

Solution #1. The most radical solution is to buy a cheap home distiller which uses fan for cooling and has productivity of about 1 liter per 1.5-2 hours. The latest distiller I have found on Internet is shown on Figure 1.1. I used the Milli-Q water for purification. Some institutions have pipelines with distilled water.

The price of the distiller (Fig. 1.1) in 2011 was $165 online. This distiller produces 4 gallons of pure water per working day. Manual filling is easy and safe. Included with distiller are all parts and supplies to get started: collection bottle, cords, filters, and residue cleaner, easy to follow directions. 4 gallons of water is more than enough for 2-3 months of experiments with two isolations a week. I think that this is an optimal performance because you usually have to spend a day for working with results, making figures and planning the next experiment.

I found that it is not productive doing experiments day after day without analyzing previous experiments and knowing where you are, and finding possible errors early. For each day I make plans, what experiments I have to do and what to expect.

Solution # 2. It is also acceptable to add 50-100 μl of catalase suspension (from Roche, Cat. No. 106 810) per 1 gallon of water. After preparation of a buffer, I filter sterilized the solutions using cellulose acetate 0.22 μm filters from Whatman (Zap Cap-S). These filters are less liable to get clogged with sucrose, and thus allow filtering 2-4 liters of incubation buffer with just 1 or 2 filters. Because during sterilization of solutions by filtering through membrane the amount of hydrogen peroxide may also increase, I also added 50-100 μl of the catalase suspension to 2-4 liters of the isolation buffer, or correspondingly less to the incubation buffer. Interestingly, in the control experiments I found that the presence of catalase in the incubation medium did not affect the rates of ROS production, measured as H_2O_2 by the Amplex red method. Evidently horse radish peroxidase and Amplex red work much faster than catalase.

Solution # 3. If you suspect that water might be contaminated with Fe ions, you can add 1 or 2 teaspoons of Helix-100 to 0.5 - 4 liters of isolation or incubation buffer and stir for about 15-20 minutes. However, Helix-100 must be added to the medium before addition of EGTA. After that, you filter sterilize you buffer, add EGTA and adjust pH.

Some other remarks regarding the effects of substrates and other chemicals on fluorescence of solutions and measurements of ROS production will be discussed in the section describing measurements of ROS production by mitochondria.

The Purity and Quality of Chemicals.

General remarks. The purity of chemicals is very important for the mitochondrial studies. Practically all chemicals have to be of the top purity and quality. After having some unexpected problems, which originated from the quality of the chemicals, I started to use the most expensive brands, usually of the Molecular Biology quality. Once, while I was at Emory, I asked our Lab Manager to buy sucrose for my experiments. Being a Manager, she selected from the Sigma's catalog the less expensive brand of sucrose. As a result, for several days my control mitochondria were very bad. The problem was resolved after I asked the Manager to buy the top quality sucrose. When I studied transport of calcium into mitochondria using Calcium Green dye (Molecular Probes), I quickly realized that many chemicals, including the high quality sucrose and particular mannitol, were highly contaminated with calcium. I had to use EGTA in order to have the background fluorescence in the presence of Calcium Green at the minimal level. Some authors recommended using the isolation medium without EGTA during the final spinning down of mitochondria. I tried this and found that the background fluorescence with Calcium Green was beyond the scale. Therefore, I used the isolation medium with EGTA for the final sedimentation as well, but made the working suspension of mitochondria in a medium without EGTA. The remaining EGTA in the mitochondrial pellet was sufficient to hold the background fluorescence at reasonably low level.

In practice, if I have found that this particular bottle or vial of a chemical works fine in my experiments (KCl, phosphate salts, and other chemicals) I kept this bottle close to myself and used it for many years until it lasts. In comparison with other types of research, say cell culturing or immunohistochemical methods (Western, Northern blots, etc), mitochondrial studies are relatively cheap. "Classical" Mitochondriology requires much less expensive chemicals. Therefore, some of the chemicals may be stored for decades, and it is not always true that a newly bought chemical will be better than the one, which was on your shelf for two decades. Once, I helped to establish some of my methods to a colleague of mine in London. She invited me to her Lab and paid the travel expenses. I taught her how to make the TPP-sensitive electrodes for measurements of the membrane potential. I brought with me a bottle of tetraphenylboron from Aldrich, which I used for more than 10 years. I left this bottle with my colleague thinking that, when I will be back in USA, I will buy a new bottle of the chemical. I was wrong! It cost me so many troubles! Before, when I used the old batch of tetraphenylboron for preparation of the TPP-sensitive membrane, the electrodes would work for many months, even for years. When I bought a new batch of the chemical from Sigma, the electrode would work no longer than for 2-3 days. After that, the sensitivity would drop significantly, and the electrode would become of no use. I tried several samples of the chemical, made many membranes and electrodes and still, the results would be unacceptable because of low sensitivity and reproducibility. Only when I bought a new vial of tetraphenylboron from Aldrich, the electrodes become somewhat better, but finally, I had to ask my colleague to share with me the chemical I left in her lab.

I hate to mention it, but the quality of many chemicals from Sigma is poor. Though, this regards not only to Sigma. If you do some sensitive quantitative experiments, such as estimation of the ADP/O ratios, Michaelis constants (K_M) for substrates, ADP, NAD or NADH, and such, you have to check the quality of the chemicals using HPLC or the enzymatic methods. For example, when I studied the effects of adenine nucleotides on the adenine nucleotide translocase (see Panov et al, 1980) it was essential that the chemicals would contain only ADP, ATP or AMP. So, I used HLPC to test the purity of my chemicals. I found that some brands of ADP from Sigma or Calbiochem would contain up to 45% of AMP, 15% of adenosine; some vials of ATP would contain 30% ADP, 15% AMP. In one case, the bottle from Calbiochem, marked as ADP, was 100% AMP. Therefore, I usually avoid measuring ADP/O ratios because you have to be sure that your sample of ADP does not contain AMP and adenosine.

Some chemicals require additional purification. For example, Ruthenium Red (from Aldrich) would not work as an inhibitor of Ca^{2+} transport without additional purification. The purified Ruthenium Red works very well and you can use it for years. The method of RR purification is described in the Supplements.

When studying production of ROS by mitochondria, and using various inhibitors and uncouplers for analysis of the sites of ROS production, a big problem is finding a sample of ethanol without background fluorescence. To test the quality, just add ethanol alone in a quantity you use for inhibitors in the presence of Amplex red and HRP, and check changes in the fluorescence. Most samples of ethanol would give a huge jump in fluorescence. Evidently, at some point this ethanol was stored or exposed to plastic. When you find a batch of good ethanol (I used ethanol for HPLC, though it was denatured with isobuthanol), keep that bottle close to yourself for the Future. Ethanol from plastic bottles usually has high fluorescence. Unfortunately, the chemical companies more and more often use plastic containers for ethanol.

Major chemicals used for isolation and incubation Buffers

Sucrose. Sucrose, or table sugar, is a dimer of glucose (left) and fructose (right). Molecular mass 342.3 g/mol, density = 1.588. The solubility of sucrose is high – 200 g/100 ml. Therefore it is easy to vary densities of the isolation buffers for various purposes. It is the major component of the buffers for isolation of mitochondria. Sucrose does not penetrate the inner membrane of mitochondria.

Mannitol. Mannitol is a sugar alcohol with a molecular mass of 182.17 g/mol, density = 1.489. It is poorly metabolized and by 90% is excreted via kidneys.

Mannitol is almost 10 times less soluble than sucrose, 22 g/100 ml. Mannitol is also osmotically active. In the literature the isolation buffers for preparation of the heart and brain mitochondria and other mitochondria researchers use 225 mM Mannitol and 75 mM sucrose. It is impossible now to find out why this composition was recommended. The serious drawback of this medium with high concentration of mannitol is that mannitol enters the red blood cells, and therefore the swollen erythrocytes are not easily removed from the homogenate during the 1st centrifugation. As a result, mitochondria become contaminated with erythrocytes. This is the reason why liver mitochondria are isolated in the isolation buffer, which contains sucrose as the only density component. I have also found that in comparison with sucrose, mannitol is much more heavily contaminated with calcium and possibly with other chemicals, possibly because of the nature of its production (see Wikipedia). For these reasons, I reversed in my isolation buffers the contents of mannitol and sucrose: 225 mM sucrose, 75 mM mannitol. The quality of my brain and heart mitochondria was good and contamination with erythrocytes was much lower. Though, I always used the Percoll gradient with a layer of 40% Percoll to eliminate erythrocytes completely. Erythrocytes contain many enzymes, which may interfere with your measurements. In addition, the density of the isolation medium with high mannitol is significantly lower than with high sucrose. I used to add mannitol because it indeed dimishes the oxidative damages to mitochondria during experiments with the diseased animals.

Buffers

MOPS. This buffer is the common name for the compound 3-(N-morpholino) propanesulfonic acid, a buffer introduced by Good et al. in the 1960s. Its chemical structure contains a morpholine ring. HEPES is a similar pH buffering compound that contains a piperazine ring. With a pKa of 7.20, MOPS is an excellent buffer for many biological systems at near-neutral pH. I use usually this buffer for isolation and incubation buffers because in the cytosol the physiological pH is 7.2. Thus with 10 mM MOPS you will have much higher buffering capacity in the pH range 7.2-7.4 than with 10 mM Tris-HCl, which has pKa of 8.06.

Tris-HCl has a pKa of 8.06 at 25°C , which implies that the buffer has an effective pH range between 7.1 and 9.0. That is at pH 7.2 the buffer capacity of Tris is low. Since I used AgCl reference electrodes for measuring pH and membrane potential, I eliminated Tris because it is incompatible with the silver-containing combination pH electrodes. Tris forms insoluble complex Ag-Tris, which clogs the junction. Common forms of Tris are Tris-HCl, which is the acid salt, and Tris-Base, which is alkaline. When titrated to pH = pKa with the corresponding counterion (OH- for tris-HCl, H+ for tris base) they have equivalent concentrations. However, the molecular weights are different and must be correctly accounted for in order to arrive at the expected buffer strength.

HEPES (4-(2-hydroxyethyl)-1-piperazineethane-sulfonic acid) is a zwitterionic organic chemical buffering agent. HEPES is widely used in cell cultures largely because it is better at maintaining physiological pH despite changes in carbon dioxide concentration (produced by cellular respiration), when compared to bicarbonate buffers, which are also commonly used in cell cultures. HEPES has two pKa points at pH 3.0 and 7.55. The dissociation of water decreases with falling temperature, but the dissociation constants (pK) of many other buffers do not change much with temperature. HEPES is like water in that its dissociation decreases as the temperature decreases. This makes HEPES a more effective buffering agent for maintaining enzyme structure and function at low temperatures. Lepe-Zuniga et al. (1987) reported a phototoxicity of HEPES when exposed to ambient light by the production of hydrogen peroxide, which is not a problem in bicarbonate-based cell culture buffers. It is therefore strongly advised to keep HEPES-containing solutions in darkness as much as possible. And HEPES is not recommended if you study production of ROS by mitochondria.

Chelators

EGTA (ethylene glycoltetraacetic acid) is a polyamino carboxylic acid, chelating agent that is related to the better known EDTA, but with a much higher affinity for calcium than for magnesium ions. EGTA has 10000 times lower binding affinity for Mg^{2+} than for Ca^{2+}. It is useful for making buffer solutions that resemble the environment inside living cells where calcium ions are usually at least a thousandfold less concentrated than magnesium. The pKa for binding of calcium ions by tetrabasic EGTA is 11.00, but the protonated forms do not significantly contribute to binding, so at pH 7, the apparent pKa becomes 6.91.

To obtain different concentrations of free Ca^{2+} in the incubation buffers it is practical to use the Ca^{2+}/EGTA buffers. The concentration of free Ca^{2+} below 5 – 10 µM can be measured using Fura 2 dye (Molecular probes). At the ratio of 0.7: 0.7 mM $CaCl_2$ and 1 mM EGTA, the concentration of free Ca^{2+} is close to 1 µM (my own data). There is a software, which allows calculating $[Ca^{2+}]_{free}$ at different concentrations of Ca^{2+} and EGTA (http://maxchelator.stanford.edu/CaMgATPEGTA-TS-Plot.htm).

Because during homogenization of a tissue large amount of calcium is released, it is necessary to have in the isolation buffer 1 mM

EGTA. In earlier days, when researchers did not used EGTA in the isolation buffers, the mitochondria were usually overloaded with calcium. See my paper (Panov & Scarpa. *Biochemistry* (USA), 1996, 35, 12849-56). EGTA does not affect significantly the concentrations of Mg^{2+}, Cu^{2+} or Zn^{2+} ions present in water or buffer.

EDTA (Ethylenediaminetetraacetic acid) is a polyamino carboxylic acid well soluble in water. Its conjugate base is named ethylenediaminetetraacetate. Its usefulness arises because of its role as a hexadentate ("six-toothed") ligand and chelating agent, i.e. its ability to "sequester" metal ions such as Ca^{2+} and Fe^{3+}. After being bound by EDTA, metal ions remain in solution but exhibit diminished reactivity. EDTA is produced as several salts, notably disodium EDTA and calcium disodium EDTA.

EDTA has several orders higher affinity than EGTA in binding Fe, Mg, Cu and Zn ions. Therefore, when Milli-Q water becomes toxic due to the presence of Cu and Zn, the toxicity can be eliminated by EDTA but not by EGTA. Researchers often use EDTA indiscriminately. As a result, the functions of mitochondria may change dramatically. I will discuss in details the effects of EDTA on mitochondrial respiration and ROS generation in Chapter 7 describing incubation media to study mitochondrial functions.

The major side effects of EDTA on mitochondrial functions are bound to the removal of Mg^{2+} from the high affinity Mg-binding sites in mitochondria. This results in conformational changes in some proteins and increased conductivity of the inner membrane for protons. In addition, removal of Mg^{2+} from the medium changes the kinetics of interactions of the mitochondrial enzymes with ATP and ADP. Normally, almost all ATP in the cells is bound to Mg^{2+}.

BSA – Bovine Serum Albumin. Many authors use defatted bovine serum albumin (BSA) in the isolation medium. The detailed discussion of the BSA usage will be given later, when I will discuss the succinate-dependent reversed electron transport and experiments on animals with modeled diseases.

Some researchers use up to 0.5% of BSA (weight to volume), that is 5 grams of BSA per 1 liter of the Buffer. That is too much. I use no more than 1 gram of BSA per 1 liter that is 0.1%. You may find a wide variety of BSA usage for isolation of mitochondria. Some authors isolate mitochondria without BSA and add it to the final suspension of mitochondria; other authors add BSA to the incubation medium. Many authors, including myself, use BSA added to the isolation buffer from the beginning. Although I have never seen a discussion of what brand of BSA is preferable and why, I made some observations from my own experience.

It is believed that BSA protects mitochondria from uncoupling by fatty acids, which are released during homogenization of a tissue. This belief exists since late 60s. But this is not exactly true. In our work (Panov et al., 2010), we have shown that BSA may remove oxaloacetate from the binding sites on mitochondria and thus increase the activity of succinate dehydrogenase. Brain mitochondria isolated without BSA present in the isolation medium may not oxidize succinate and even generate membrane potential. Therefore, many researchers regarded these mitochondria as "bad", and those isolated with BSA as "good". These effects of BSA on succinate oxidation by brain mitochondria are subject to variation between species, and organs. Whether BSA is present in the isolation medium or not, has no effect on the liver and heart mitochondria of Sprague Dawley rats (Panov et al., 2010).

What is also important, that BSA may protect mitochondria from oxidative damages caused by the contamination of the isolation medium with hydrogen peroxide. This is particularly important when isolating brain or spinal cord mitochondria from the SOD1 transgenic animals. Mitochondria from diseased animals are particularly vulnerable to oxidative stress. Therefore, I always used BSA when working with animal models of diseases. In my experiments with the transgenic mice (YAC72) expressing human mutated gene encoding huntingtin (Htt) protein, I have found that when liver or brain mitochondria were isolated in the presence of BSA, the mitochondrial functions were perfectly normal. I observed the specific mitochondrial dysfunctions only in mitochondria isolated without BSA present in the isolation or incubation buffer.

There are many brands of BSA in the Sigma's catalog. Serum albumin may be referred to as Cohn Fraction V. This naming convention is taken from the original Cohn method of fractionating serum proteins using cold ethanol precipitation. Serum albumin was found in the fifth ethanol fraction using Cohn's method. Since then, the term "Cohn Fraction V" has been used by some to describe serum albumin regardless of the method of preparation. Others used this term to describe serum albumin purified by ethanol fractionation methods that have been highly modified since the original Cohn method was described. Sigma-Aldrich manufactures and distributes serum albumins purified from a variety of primary methods including the true Cohn fractionation method, modified ethanol fractionation methods, heat shock and chromatography. Additional purification steps may include crystallization or charcoal filtration.

I used the brand Sigma A4503-50G. This brand has been purified by treatment with charcoal that removes excess of fatty acids and other organic compounds that have relatively low affinity for binding on BSA.

The brand A6003-25G is a lyophilized powder, essentially fatty acid free, ≥96%. This brand was treated with acetone to remove fatty acids. It is two times more expensive than A4503 brand.

I used several times this expensive defatted brand, but then stopped because some of the results were different from my earlier experience. After some consideration, I seased using this expensive brand for the following reasons: BSA has many sites for binding different compounds, both organic and inorganic. Treatment with acetone made BSA more capable of binding hydrophobic compounds, including phospholipids, hormones, etc. I studied for my Ph.D. thesis the effects of aldosterone and corticosterone on oxidative phosphorylation, and have shown that these hormones are effective at very low concentrations. I concluded that acetone-defatted BSA may have some undesirable effects on mitochondria because it may remove from mitochondria important hydrophobic molecules and peptides. Therefore, I would use this brand (A6003) only for very specific tasks. Because the other brand (A4503) showed significant protective effect against oxidative stress and effectively removed oxaloacetate from mitochondria (Panov et al., 2010) I used this brand in my experiments for isolation buffers at 0.1%.

As regards adding BSA to the incubation medium, you should know that BSA significantly diminishes the sensitivity of the Clark electrode (see. Panov et al. 2003, Arch. Biochem. Biophys. 410, 1-6).

CHAPTER 2

Preparation of Buffers for Isolation of Mitochondria

Mnemonic formulas to calculate how many mg of a chemical per 100 ml you have to take in order to get a particular mM concentration; and what concentration in mM will be if you take X amount of a chemical per 100 ml:

$$\textbf{mg/100 ml} = (\text{mM x MW})/10;$$
$$\textbf{mM/L} = (10 \text{ x mg/100 ml})/\text{MW}$$

It is practical to prepare the isolation buffers in a relatively large quantity. I usually prepared 4 liters. Earlier, I have described that water should contain hydrogen peroxide as little as possible. To check the contents of H_2O_2 in your water use the amplex red method. If you do not have a wheel filter fluorometer, but a conventional spectrofluorometer, zero the instrument against water treated with catalase. The chemicals should be of highest purity available.

Table 2.1 Basic buffer for isolation of mitochondria from brain, heart, skeletal muscle and kidney.

Components	Mol. Weight	1 Liter
225 mM Mannitol	182.2	40.995 g
75 mM Sucrose	342.3	25.673 g
10 mM MOPS, pH 7.2	209.3	2.093 g
10 mM Tris	121.1	1.211 g
1 mM EGTA (Ca-free)#	380.4	380.4 mg
If present: BSA Σ A4503	66,000.00	1 g

The exact composition of the isolation buffer may depend on a particular task. Table 2.1 shows the composition of the buffer I used for heart mitochondria, which contained besides EGTA also 0.5 mM EDTA. The side effects of EDTA will be described later. This particular buffer was used by many authors and was originally also used in the Russell Scaduto's lab for isolation of the rat heart mitochondria. Because the presence of EDTA may lead to serious complications, I skipped EDTA completely, and used it only for specific purposes, for example to remove the possible excess of divalent cations (Cu, Zn) and used it only during homogenization stage when isolating the heart or skeletal muscle mitochondria. Though, most complications can be avoided by addition to the buffer of 0.5 mM MgCl$_2$. After the first sedimentation of mitochondria I used the isolation medium without EDTA.

Basically, the same isolation buffer can be used for preparation of the skeletal muscle and neuronal (brain, spinal cord) mitochondria. I used it also for isolation of the liver mitochondria, when I compared liver and other types of mitochondria (from heart or brain). Many authors use for isolation of liver mitochondria just the sucrose-based medium (see below).

In the literature you will find that most researchers use the above concentrations of mannitol – 225 mM and sucrose – 75 mM. The drawback of high mannitol is that mannitol is accumulated in erythrocytes and causes them to swell. Therefore, during spinning down the tissue's debris the supernatant fraction containing mitochondria contains also large amounts of red blood cells. In this case you MUST purify mitochondria using Percoll gradient with the lowest layer of 40% Percoll. I found it useful to reverse the concentration of mannitol and sucrose, which made the medium with slightly lower tonicity but significanbtly higher density, which allowed better separation of mitochondria.

Table 2.2. "Improved" isolation buffer. This buffer has reversed mannitol/sucrose ratio.

Components	Mol. Weight	1 Liter
75 mM Mannitol	182.2	13.67 g
175 mM Sucrose	342.3	59.9 g
10 mM MOPS, pH 7.2	209.3	2.093 g
1 mM EGTA (Ca-free)#	380.4	380.4 mg
0.1% BSA Fraction V, Σ A4503-50G	$\approx 66,000$	1 g

This medium noticeably diminishes contamination of mitochondria with erythrocytes because at lower concentration of mannitol erythrocytes do not swell much, and more red cells are sedimented with the tissue's debris. I used this medium for isolation of mitochondria from all tissues, including liver. I would miss BSA in many experiments, depending on the task. Briefly, BSA eliminates the intrinsic inhibition of succinate dehydrogenase (SDH) by endogenous oxaloacetate, which strongly depends on the metabolic phenotype of an animal. On the other hand, BSA protects mitochondria during the isolation procedure from oxidative damage, particularly, if the medium is contaminated with peroxides. Therefore, mitochondria from the experimental animals of toxic or genetic models of diseases might become dysfunctional when isolated without the presence of BSA. When you study the metabolic features of mitochondria, always remember that one of the important points of regulation of metabolism is succinate dehydrogenase (SDH). The intrinsic inhibition of varies between organs of the same animal. As well as between species. Also, BSA may remove hormones and peptides that participate I regulation of mitochondrial functions. So, be careful with BSA.

CHAPTER 3

Procedures for Isolation of Mitochondria from various Tissues and Cultured Cells

Methods to sacrifice experimental animals

The quality of mitochondria, which you intend to isolate, to a large degree will depend on how you sacrifice your experimental animal. First, you have to find a balance between the scientific goals and preservation of the humane attitude to animals. You always have to remember that animals are as much God's creatures as you are, and if you have to kill an animal, you must do it as quickly and painless as possible. In my carrier I witnessed different attitudes to this problem: from complete negligence, when the head of a large rat was cut off with dull scissors, to stupidly "humane", when the animals were forced to slowly suffocate in CO_2, which the animal rights advocates (and bureaucrats) consider as appropriately "humane". In Britain, researchers are not allowed even to touch animals. Everything is done by the "specially trained technicians", behind the closed doors, and you will be given the organ you requested through a window in the door. I remember, once working in London, I and my colleague were planning to isolate liver mitochondria. The technician gave us the rat's liver almost black, overfilled with the dark blood. When I asked the technician how he sacrificed the animal, he answered, as it was appropriate to the protocol: First, the rat suffocated for several minutes in a can with CO_2, and then he (technician) removed the liver. From the Scientific point of view, this was, of course, completely inappropriate and waste of the animal's Life!

Suffocation of rats and mice in CO_2 was (in 2002) also the usual protocol at the University of California (in Los Angeles). As a result, the organs for several minutes were subjected to hypoxia and even ischemia.

At other institutions the animals were first anesthetized with Isoflurane (2-chloro-2-(difluoromethoxy)-1,1,1-trifluoro-ethane, a halogenated ether used for inhalational anesthesia), then decapitated with guillotine, and the body was drained of blood before removal of the liver. As the anesthesia becomes complete: the rat's body is fully relaxed, the rat is placed into a special plastic cone head first, which immobilizes the animal. The head of the animal, to be sacrificed, is inserted into the instrument's orifice in such a way, that the blade of the guillotine will be just behind the ears and the body is tilted (head down) by about 35-40 degrees. The handle is pressed down sharply. In this way the head will be cut off along the back of the skull with minimal damage to cervical bones of the spinal cord and minimum muscles attached. The foramen magnum will be immediately visible. This guillotine is rather expensive: about $450.00 (Fisher Catalog, 2009). Isoflurane, however, may affect the experimental results because it is highly soluble in lipids.

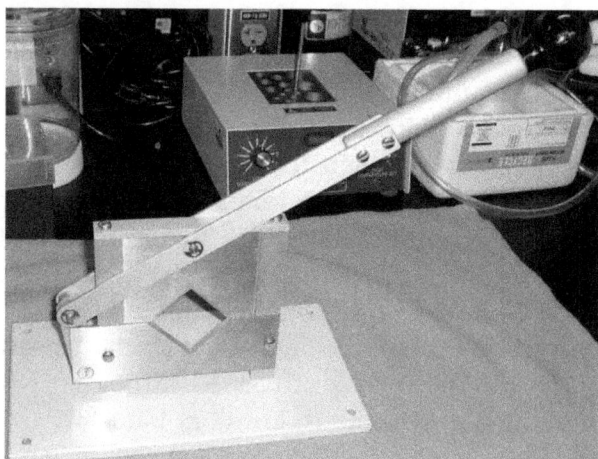

Figure 3.1. Guillotine for small animals (rats).

Therefore, if you study some activities of the cytochrome p-450 in liver, you have to keep this in mind. I closely watched the animal's behavior during anesthesia, and as soon as the animal became limp, I removed the animal from the can with the anesthetic and decapitated as soon as possible (ASAP). To my experience, Isoflurane did not affected the quality of the brain mitochondria.

Rat can also be anesthetized by a mixture of Ketamine (50 mg/kg) plus Xylazine ((0.5 mg/kg). Ketamine is primarily used for the induction and maintenance of general anesthesia, usually in combination with a sedative, such as Xylazine. But this type of anesthesia is inappropriate if you plan to work with cytochrome P-450.

Mice are small animals. Therefore, the usual procedure is to kill the animal with the head dislocation by large scissors – you have to do it really fast, and cutting off the head. But you can also use Isoflurane for quick anesthesia.

Usually the organs removed from the body are placed into a beaker filled with a mixture of liquid and frozen isolation buffer (ice-slurry cold). Ice should constitute at least 30% of the volume.

Commentary: You will get significantly higher yields of mitochondria per 1 g of tissue from mice as compared to rats, and usually mice mitochondria have higher respiratory activities.

Methods of tissue disintegration and homogenization.

Disintegration of soft tissues such as brain, spinal cord and cultured cells. Although there are different sizes of Dounce grinders, the optimum size is 40 ml. The grinder has two ball pestles: the tight one, and the loose one. It can be used to isolate mitochondria from the brain and spinal cord, and from the cells. With brain and spinal cord I used only the "loose" pestle. With the cells, I used both pestles: initially the "loose" one, and then the "tight" pestle. The trick is to create a "vacuum" by explosive movement of the pestle up. After that, you slowly push the pestle to the bottom and then repeat the sharp movement up. You must wear latex gloves, and with the left hand cover the top of the glass tube preventing spilling and restricting the sharp upward movements of the right hand. It takes some time to get the experience to do it right.

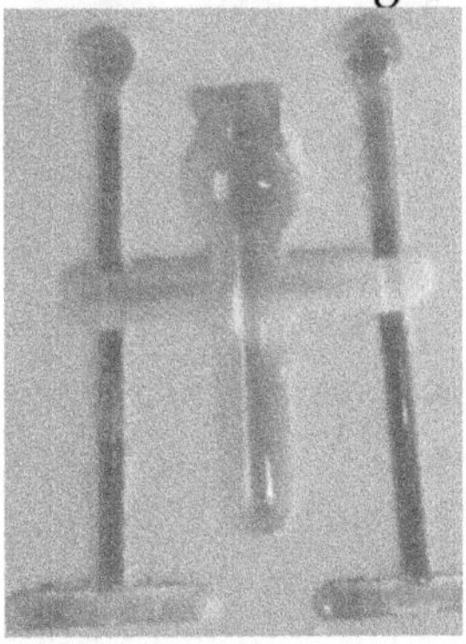

Figure 3.2. Wheaton Dounce Tissue Grinder.

The Wheaton Dounce grinder will not work well with harder tissues, such as heart, kidney and even pig liver.

Disintegration of the liver and harder tissues.

The most commonly used homogenizer is the classical Glass/Teflon grinder, which has a tube made of glass, often covered with plastic for protection against breakage, and the pestle made of Teflon. Both the bottom of the glass tube and the pestle must be completely congruent. There are homogenizers with the rounded or conical bottom. It is important that the pestle should not be too tight. The pestle should fall freely but slowly to the bottom when released. I found it optimal that the clearance between the pestle and the tube's wall is about 0.3-0.5 mm (the other side of the pestle is touching the wall of the tube). The figure below shows the classical glass/Teflon homogenizer, which has the volume (from the bottom to the rounded part) of about 40 ml. This size is practical for most purposes. Avoid loading too much tissue because it is not efficient. For approximation of the optimal amount of the tissue: one lobe of the rat liver, or one mouse liver for one homogenization in a 40 ml glass tube with 30 ml of the buffer.

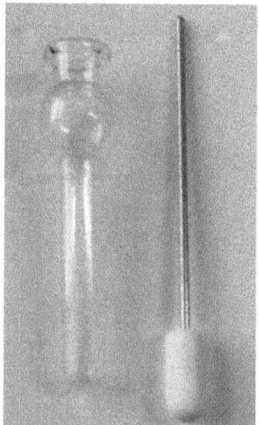

Figure 3.3. Glass 40 ml homogenizer with Teflon pestle.

For correct homogenization, it is important to use a motor, which is powerful enough to overcome the resistance during homogenization of the hard tissue, and allows regulating of rotation at relatively low speeds. I used the powerful motors from the Fisher's catalog designed for stirring of viscous solutions. These motors are expensive (in the range of $1000.00), but perfect for the purpose and will last for decades. This motor is easy to fix to a shelf in a cold room. The optimal rotation rates are 300 – 400 rpm. Both the glass tube and the pestle have to be chilled in a bucket filled with ice. Basically, all procedures have to be performed in a cold room. Before homogenization, the tissue has to be minced into small pieces by scissors (liver, brain) or with scissors and Polytron homogenizer (heart, kidney, skeletal muscle).

Figure 3.4. BioSpec 985370-14 Tissue-Tearor Homogenizer. For 5-1000mL Samples, 5000 - 35000 rpm, 120 VAC, 13.2cm Probe Length. $859.00 (in 2012).

For hard tissues and muscles, it is much faster and more efficient to use Polytron and similar tissue grinders (Fig. 3.4). Still, the tissue (skeletal muscle or heart, kidney) has to be minced into small pieces scissors. The more tissue, the larger the diameter of the Polytron's tube. Chill the tube in ice. For this, I used 25–50 ml cut in half plastic culture pipettes (depending on the tube's diameter) and stuck the tube into ice. For disintegration, place the tissue in a 25 or 50 ml centrifuge tube (depending on the amount of tissue), with the tip of the Polytron's tube at the bottom, and turn on the motor just for 2-3 seconds. Once for the liver, two or three times for harder tissues with 15 s intervals to allow the tissue to settle down. Then homogenize using the glass/Teflon homogenizer with just 3-4 strokes up/down.

Isolation of rat and mouse liver mitochondria.

Most of our current knowledge about mitochondria was obtained in experiments with liver mitochondria. The major reason for this was the simple fact that just using one animal, in a short time the researcher had plenty of mitochondria for experimentation. However, the indiscriminate usage of methods suitable for liver mitochondria for mitochondria from other organs, often led to erroneous results. Take, for example, the usage of succinate + rotenone as a respiratory substrate.

Close relationships between the functions of an organ and the mitochondria.

One of the major ideas, which I will try to unveil in this Manuscript, is a seemingly "simple" idea that mitochondrial structure and functions closely relate to the structure and functions of the organ and cells from which these mitochondria originated. Although this idea seems obvious, to this day many researchers think that there is no principal difference between brain, heart or liver mitochondria. And there is a belief, that liver mitochondria are so simple. In fact, after I worked for many years with mitochondria from different organs and cell types, I came to conclusion that of all mitochondria, the liver mitochondria are the most difficult to study. They are in many ways unique and distinct from say, heart, brain or kidney mitochondria. And this is because metabolically and functionally liver stands apart from the energy metabolism of any other organ.

The methods of isolation and measuring the respiratory activities of liver mitochondria were generally considered as simple and fundamental. However, looking back at the old papers, published during 60s and 70s, you can see that the quality of mitochondria isolated even at the leading laboratories was usually not good. There were many pitfalls unknown at the time, which I will explain gradually in the course of this Manual. In this chapter I will share my experiences and thoughts regarding the liver mitochondria.

The metabolic properties of liver mitochondria strongly depend on the metabolic state of the liver.

This is in contrast to the heart or brain, where properties of mitochondria do not depend so much on the metabolic state of the body or liver.

According to the British Liver Trust, liver has more than 500 functions. The most important of them are:
- Processing digested food from the intestine;
- Controlling levels of fats, amino acids and glucose in the blood, and thus maintaining the metabolic (substrate) homeostasis for other organs and tissues;
- Neutralizing and destroying drugs and toxins;

- Making enzymes and proteins, which are responsible for most chemical reactions in the body, for example those involved in blood clotting and repair of damaged tissues, blood serum albumin;
- Manufacturing bile;
- Breaking down food and turning it into energy;
- Storing iron, vitamins and other essential chemicals;
- Manufacturing, breaking down and regulating numerous hormones including sex hormones;
- Production of large quantities of purine nucleotides for bone marrow and neuronal tissue;
- Combating infections in the body;
- Clearing the blood of particles and infections including bacteria; etc.

In order to maintain in the blood the metabolic homeostasis for other organs and tissues, the liver must convert one type of metabolites into other metabolites. For example, carbohydrates can be converted into fats and amino acids; fatty acids and amino acids can be converted into glucose. There are several metabolic states of the liver, the appearance of which depends on the type of food and the time passed after food consumption. Of course, the metabolic properties of the liver will depend also what type of animal you study: herbivorous (rabbits), carnivorous, or omnivorous (human, rats, mice and pigs). It should be also kept in mind that everything, what enters the blood from the stomach and gut, goes directly into the liver through the portal vein. In an animal, such as a rat or a mouse, we can distinguish metabolically several states, depending on the time passed after the last food consumption: i) the absorption period, ii) the fed state, and iii) the starved metabolic state, which have different degrees of metabolic changes depending on the longevity of fasting.

1. The Absorption period. During absorption period the metabolites (carbohydrates, fatty acids, amino acids), derived by digestion of food, are delivered through the portal vein to the liver cells. Fatty acids become activated to acyl-CoAs, carbohydrates are phosphorylated, and amino acids become converted into proteins or other metabolites. In rodents (rats & mice), the liver mitochondria are capable to transport and activate fatty acids inside mitochondria in the matrix without participation of carnitine. Activation of fatty acids results in formation of adenosine monophosphate:

Fatty acyl + ATP + CoA → Fatty-acyl-CoA + AMP

Thus, AMP can be accumulated in the cytosol and the mitochondrial matrix. Therefore, if the liver mitochondria were isolated within 1-2 hr post feeding of the animal, the rate of oxidative phosphorylation will be very low because of the low content in the matrix of exchangeable mitochondrial adenine nucleotides ([ADP] + [ATP]). We have shown that liver mitochondria isolated during absorption period contain large amount of AMP. The AMP content in the mitochondrial matrix strongly depends on the type of the diet. If the food was rich in fats, the level of AMP will be high. Because long-chain acyl-CoAs are powerful inhibitors of adenine nucleotide translocase (ANT), oxidative phosphorylation will be inhibited and the hepatocyte's [ATP]/[ADP] and the phosphate potential ([ATP]/[ADP] + [Pi]) ratios also will be low.

Figure 3.5A shows that liver mitochondria (LM) from the fed rats at the absorption period showed a remarkable inhibition of succinate oxidation both in the state 4 (resting respiration) and state 3 (phosphorylating respiration). However, upon addition of uncoupler the state 3U was similar to that of LM from the fasted rats. Consecutive additions of ADP gradually increased the rates of the state 3 respiration. But even after the 5th addition, the state 3 rate remained inhibited by 40-50% as compared with the state 3U. Both carnitine (Fig. 3.5B) and BSA (Fig. 3.5C) had no effect on the state 3 respiration of LM from rats at the absorbtion period. This is because limitation of the state 3 respiration was associated with the low exchangeable pool of adenine nucleotide in the matrix ([ADP] + [ATP]) due to high concentration of AMP ([AMP]).

In the experiment *in vitro*, the rate of oxidative phosphorylation can be restored by incubation of mitochondria with α-ketoglutarate, which produces ATP during the substrate phosphorylation step and restores the exchangeable pool of adenine nucleotides in the transphosphorylation reaction: ATP + AMP → 2ADP. (Panov A.V., Konstantinov Yu.M., Lyakhovich V.V. The possible role of palmitoyl-CoA in the regulation of the adenine nucleotide transport in mitochondria under different metabolic states. J. Bioenergetics.,7 (1975) 75-85).

2. The Fed metabolic state. 2-3 hours after feeding, the isolated liver mitochondria showed the highest rates of oxidative phosphorylation, low State 4 respiration and therefore the highest respiratory control ratios (RCR). This is the so called "fed" metabolic state of the liver. In this metabolic state the liver had high content of carbohydrates, therefore the pyruvate dehydrogenase complex was not inhibited, and the level of long-chain acyl-CoAs esters was low (Panov et al., 1975, J. Bioenergetics, 7, 75-85). In the "fed" metabolic state, the liver mitochondria readily oxidized pyruvate + malate, citrate, isocitrate and α-ketoglutarate, as well as glutamate, or succinate.

The longevity of the "fed" metabolic state is different for different species of rats (Panov et al., 1991, Int. J. Biochem. 23 (9), 875-879), as well as between rats and mice. The smaller the animal, the shorter is the fed metabolic state. For different strains of rats the longevity of the fed state may vary from 4 to 12 hours passed from the time of the last feeding (Panov et al., 1991). The time of transition from the fed to the starved metabolic state was characteristic for each strain of rats. In mice, the transition is much faster than in rats, because they are much smaller animals (the Kleiber's Law). Though, I never estimated the exact transition times for mice.

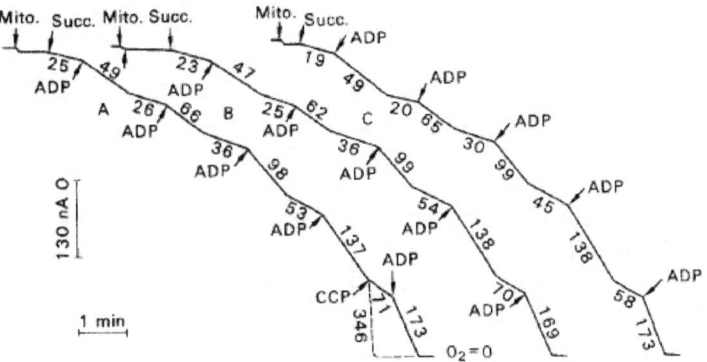

Figure 3.5. Influence of carnitine and BSA on respiration of mitochondria from the livers of rats 1.0 hour after feeding (absorption period). Note: in the paper this figure was erroneously indicated as the "fed" state.

The incubation medium contained: 100 mM KCl, 50 mM sucrose, 20 mM Tris-HCl (pH 7.4), 5 mM phosphate, 0.5 mM MgCl$_2$, 3 μg rotenone. Additions: 5 mM succinate, 150 μM ADP, 0.5 μM CCCP (chlorocarbonylcyanide). (A) Control; (B) 0,5 mM carnitine; (C) preincubation with 0.1% BSA. (From: Panov et al., J. Bioenergetics, vol. 5, 75-85, 1975).

Information: According to the Kleiber's Law, the metabolic rate for all organisms follows exactly the 3/4 power-law of the body mass, i.e., $q_0 = M^{3/4}$. It holds well from the smallest bacterium to the largest animal. The relation remains valid even down to the individual components of a single cell such as the mitochondrion, and the respiratory complexes (References: Wang et al. 2001; Rau 2002).

3. The Starved metabolic state. The fed state is followed by the "starved" metabolic state, when the liver switches from oxidation of carbohydrates to predominant oxidation of fatty acids and gluconeogenesis. In this metabolic state, hepatocytes accumulate long-chain acyl-CoAs, which inhibit oxidative phosphorylation and slightly uncouple mitochondria due to inhibition of the mitochondrial ANT and fixation of ANT in the "c" conformational state, which increased the conductivity for H^+ and K^+ (Panov et al., 1975, 1980).

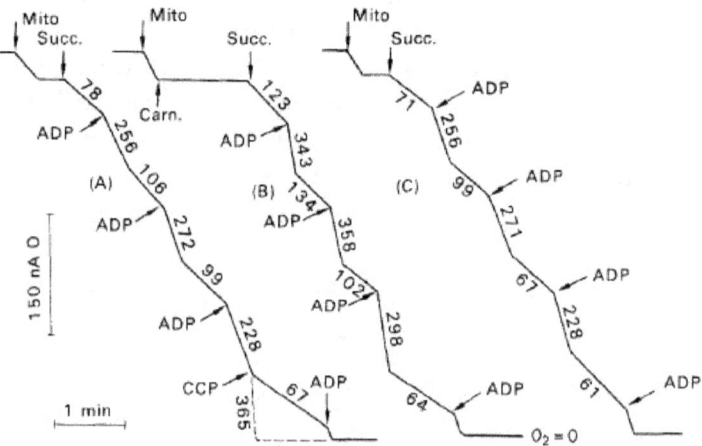

Figure 3.6. Influence of carnitine and BSA on respiration of liver mitochondria from starved (12 hrs) rats. (A) Control; (B) 0,5 mM carnitine; (C) preincubation with 0.1% BSA. (From: Panov et al., J. Bioenergetics, vol. 5, 75-85, 1975).

There are large variations in the mitochondrial activity due to variations in the longevity of the absorption and fed metabolic states, which also depend on the type of a diet and the species metabolic phenotype. Therefore most researchers isolate liver mitochondria from the animals starved overnight that is when the animals are far into the "starved" metabolic state.

In the starved metabolic state, the mitochondrial pyruvate dehydrogenase complex is inhibited, therefore the tricarboxylic acids

(TCA) cycle does not function normally, and the liver mitochondria do not oxidize pyruvate, citrate, and poorly oxidize α-ketoglutarate. The only remaining substrates for respiration are glutamate + malate, succinate and fatty-acids. To study oxidation of fatty acids by liver mitochondria is not a simple issue. Therefore, traditionally researchers, working with the liver mitochondria, used glutamate + malate or, more commonly, succinate + rotenone as the substrates. Oxidation of succinate in the absence of rotenone becomes gradually inhibited by accumulating oxaloacetate. In order to avoid this inhibition the researchers used (and still use) succinate + rotenone. This is highly unphysiological and misleading. It is much more correctly to utilize a mixture of succinate 5 mM + glutamate 5-10 mM + malate 2 mM. In this case respiration of LM from the fasted animals will be at maximum and linear in all mitochondrial metabolic states.

Figures 3.6A and 3.6B show that in the presence of carnitine there was a significant increase in the rate of the state 3 respiration. This is because carnitine removed long chain acyl-CoAs, which are powerful inhibitors of ADP/ATP carrier (ANT).

The increase of the state 4 respiration in the presence of carnitine may be explained by the fact that transport of acyl-carnitines is the energy-dependent process, but mostly because acyl-CoA dehydrogenase feeds electrons to the membrane's pool of CoQ, exactly as does Complex II during oxidation of succinate. As a result, the energy-dependent reverse electron transport and ROS production are stimulated and this increases the state 4 respiration.

Figure 3.6C shows that addition of defatted BSA into incubation medium caused a slight decrease in the state 4 respiration rate. BSA had no influence on the state 3 respiration. This indicates that inhibition of the state 3 respiration in LM from fasted rats was not caused by fatty acids, but rather by l.c. acyl-CoAs, which have much higher binding affinity to ANT than to BSA. In addition, there is evidence that fatty acids do not uncouple rat liver mitochondria (Panov et al. 2010).

I used to isolate mitochondria from the livers of rats that were sacrificed 4-6 hours after feeding that is in the fed metabolic state. For this, we controlled feeding of the animals by providing food twice a day for 1 hour at 8 AM and 8 PM. Rat liver mitochondria (RLM) from the fed rats oxidized pyruvate, and also oxidized succinate (rotenone not added) in State 4 and State 3 without noticeable inhibition by oxaloacetate (OAA). The OAA-dependent inhibition of succinate oxidation developed only after addition of an uncoupler. Subsequent addition of glutamate accelerated the State 3U until the oxygen was fully consumed.

It should be kept in mind that liver mitochondria contain less respiratory complexes as compared with the heart or brain mitochondria. In addition, for the liver mitochondria maintenance of the high ATP/ADP level in the cytosol or the high rate of ATP synthesis are not the major functions. In hepatocytes many key ATP-consuming reactions in the anabolic or anaplerotic (gluconeogenesis, synthesis of purines), and catabolic (urea synthesis) metabolic pathways occur in the mitochondrial matrix. Furthermore, liver and liver mitochondria can withstand up to 45 minutes of total ischemia if the animal was in the fed metabolic state, and 30 minutes, if the animal fasted 12 hrs before the experiment.

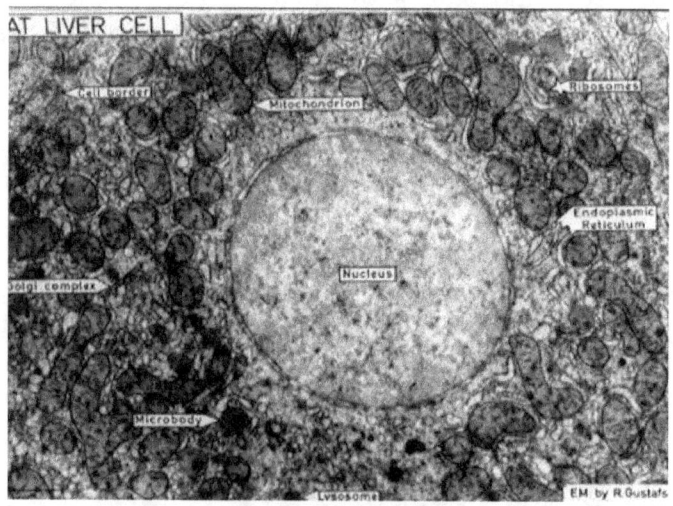

Figure 3.7. The electron microscopic picture of a liver cell. This picture was given to me in 1971 by Lars Ernster, my Scientific Father. It is a classic picture of a hepatocyte obtained by R. Gustafsson.

It should be also remembered that metabolic features of the liver cells and liver mitochondria are subjects to strong species and strain variations within the same species.

Figure 3.7 shows that in hepatocytes, mitochondria are predominantly located close to the nucleus, which reflects the high synthetic activities in the liver. For anabolic synthetic reactions, which occur in hepatocytes, the energetic value of ATP or NAD(P)H are much more important than the rates of ATP and NAD(P)H production. The energetic values are determined by the ratios: [ATP]/[ADP] ratio; [ATP]/ [ADP] + [Pi] - phosphate potential, which reflect oxidative phosphorylation; the Energy Level, or the Energy Potential: [ATP] + ½[ADP]/ [ATP] + [ADP] + [AMP], which depends on both oxidative phosphorylation and adenylate kinase reaction (Atkinson D.E. 1968, 1977, 1978).

Since liver has high regenerative potential and contains young hepatocytes with young mitochondria, as well as older mitochondria at different stages of fusion/fission, the total mitochondrial fraction of the liver can be subdivided into several fractions according to their size and density. Some authors identified 5 fractions. From the practical point of view, you have to discriminate the following 4 fractions: very heavy, heavy, middle and light fractions.

Very heavy liver mitochondria are large; probably because of fusion of several smaller mitochondria (you can see one on Fig. 3.7 in the left low corner). This fraction will be partly sedimented together with the tissue debris at 900-1000 g. The heavy fraction of large mitochondria will be sedimented between 1200 – 3000 g, the middle fraction is the largest in the quantity, will be sedimented up to 9000 g. Mitochondria that are sedimented at accelerations higher than 9000 g belong to the light fraction and contain, beside smaller mitochondria, lysosomes and peroxisomes. Lysosomes and peroxisomes, if they are present, cause mitochondria loose integrity sooner than mitochondria in the middle fraction that do not contain these subcellular particles.

Procedures for isolation of liver mitochondria

Table 3.1. The sucrose buffer for isolation of the liver mitochondria.

Components	Mol. Weight	1 Liter
250 mM Sucrose	342.3	85g 575 mg
20 mM MOPS, pH 7.2	209.3	4g 186
1 mM EGTA	380.4	380.4 mg

IMPORTANT. Rat and mouse livers are soft and can be homogenized using a glass/Teflon homogenizer. The tissue of a pig liver is very hard; therefore homogenization procedure must include preliminary treatment with Polytron, as described for the heart and kidney tissues.

In most cases liver mitochondria can be isolated using simple 0.25 M sucrose-based buffer and a glass homogenizer with a Teflon pestle. The composition of the medium is shown in the table above. This medium minimizes contamination with erythrocytes.

All procedures have to be performed (with exception of animal sacrifice) in a cold room. Prepare several containers with ice, which hold the vessel with the isolation buffer, homogenizer, beakers and the centrifuge tubes.

Step 1. A rat or a mouse MUST be sacrificed by decapitation in order to bleed the body as much as possible and prevent overfilling of the liver with blood.

Step 2. Remove the liver (if it is a mouse) or cut a piece from the lobe or the whole lobe of the liver (for a rat) and place it into a beaker containing liquid and frozen (ice-slurry) isolation buffer. The beakers are inserted into ice.

Step 3. Place the liver in the beaker with cold isolation buffer, which completely covers the tissue, and cut the tissue into small pieces with scissors; remove the liquid stained with the blood using a piece of cheesecloth, and rinse once or twice with 4-5 volumes (regarding the tissue's volume) of cold isolation buffer.

Step 4. Transfer the tissue into the glass homogenizer (40 ml volume), add 7-8 volumes of the ice-cold isolation buffer, and make 5-7 up and down strokes with the Teflon pestle rotating at 300 -400 rotations per minute (rpm). Do not try to disintegrate all pieces of the tissue. The smaller initial pieces of the liver, the higher the yield.

IMPORTANT. Do not use the tight pestle. The total distance between the glass wall and the side of a pestle has to be 0.3 – 0.5 mm. The too tight pestle and long homogenization will damage mitochondria. Some of the heavy mitochondria are really large. It is better to use more tissue and homogenize it in 2-3 portions than to homogenize the tissue for too long. If you have a lot of tissue (such as the whole rat liver), split it into 2-3 separate homogenizations. The yields are high, and the quality strongly depends on the gentle and quick handling of the tissue.

It is acceptable also to use the Polytron homogenizer. Activate the Polytron just for 2-3 seconds, holding the tip of the homogenizer at the bottom to prevent too many bubbles. After Polytron you do not need glass homogenizer. The yield of the mitochondria will be much higher as compared with the glass/Teflon homogenizer.

Step 5. Filter the homogenate through a tight mesh by squeezing the mesh between the pincer's branches. For a mesh I used a piece of my wife's Capron stockings or some bride's veil. Add more isolation buffer (usually doubling of the initial volume is enough) and split the homogenate using even number of centrifuge tubes. The diluted homogenate should not be thick. Do not overfill the tubes.

Step 6. Place the tubes into SS34 rotor of the Sorvall centrifuge and spin for 5 minutes at 900-1000 g. The rotor has to be cooled to 2-4° C.

IMPORTANT. In the earlier descriptions of the isolation procedure for liver mitochondria, it was recommended to perform the sedimentation of the tissue debris at 800 and even 600 g. For the liver this is not practical because the liver always contains large amount of blood and you have to avoid contamination of your mitochondria with erythrocytes. Therefore, I used spinning at 900-1000 g, or even higher. If you will see in your mitochondria red blood cells at the bottom, than increase the speed of the 1st centrifugation to avoid this. The yields still will be high.

Step 7. Remove the supernatant into another centrifugal tube. Pay attention that the sediment would be at the low of the tube (not aside or up) and stop as soon as you will see a "tail" from the sediment. Add new buffer, mix the sediment and equilibrate the levels of the tubes.

Step 8. Spin down the supernatant after the 1st centrifugation for 10 min at 8000 - 9000 g.

Step 9. Remove the supernatant. You can carefully turn down the tube. More safely is to remove the supernatant by vacuum suction. Remove also the lighter layer above the darker mitochondrial sediment. Resuspend the sediment in the isolation buffer. At this point you can combine the sediments from two tubes into one.

Step 10. Spin down the tubes for 10 min at 8000 – 9000 g.

Step 11. Remove carefully the supernatant and prepare the working suspension of the isolated mitochondria. For suspension of the liver mitochondria you can use 0.25 M sucrose + 10 mM MOPS, pH 7.2, or the KCl-based incubation medium. The optimum concentrations are about 20 – 30 mg/ml.

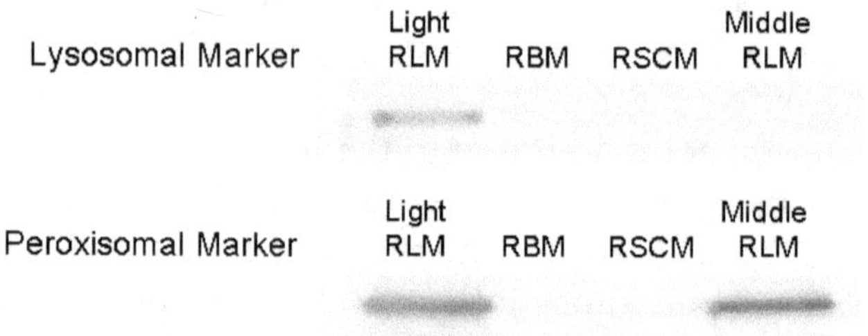

	Light			Middle
Lysosomal Marker	RLM	RBM	RSCM	RLM

	Light			Middle
Peroxisomal Marker	RLM	RBM	RSCM	RLM

Figure 3.8. Western blots of brain, spinal cord, and liver mitochondria with the marker antibodies against lysosomes and peroxisomes. The following primary antibodies were used: peroxisome marker rabbit polyclonal to catalase (Ab1877, Abcam); peroxisomal membrane protein - PMP70 (NBP-1-2071, Novas Biologicals); lysosome-associated membrane protein (Ab71489 –Santa Crus biotechnology, Inc.). As a positive control we used the liver homogenate fraction between 10,000 and 15,000 g, which contained a mixture of light mitochondria, lysosomes and peroxisomes. Because catalase was also present in the heavy and middle fractions of liver mitochondria, we used antibodies for additional peroxisomal marker PMP70 (From Panov et al., 2011. Am. J. Physiol., 300, R844-R854).

Commentary. Using this procedure you will receive the suspension of the heavy and middle fractions of the liver mitochondria, which will not contain lysosomes and peroxisomes. You should remember that intact liver mitochondria cannot be used to study production of reactive oxygen species. Figure 3.8 shows that the middle fraction of RLM (lower than 10000 g) contained no markers for lysosomes and peroxisomal markerPMP70 (not shown) although both peroxisomes and the middle fraction of RLM contained large amount of catalase. For this reason the LM are not suited for measurements of ROS production by whole mitochondria, but it is possible with the submitochondrial particles (Panov et al., 2005. J. Biol. Chem. 280: 42026-42035). On the other hand, liver mitochondria have other properties that are unique only for these mitochondria. For example, strong dependence on the metabolic state of the organ, high resistance to hypoxia, little or no influence on mitochondrial functions of bovine serum albumin, low sensitivity to changes in the composition of the medium. For example, liver mitochondria may display high respiratory rates in the sucrose medium, while brain mitochondria have significantly lower respiratory rates. Liver mitochondria have relatively high content of endogenous substrates, while brain mitochondria have no endogenous substrates.

Basic procedures for isolation of mitochondria from heart, skeletal muscles, kidney and brain

The basic principles of the method described in this section can be used for isolation and purification of mitochondria from different organs, even the liver. The related specifics will be described separately for each tissue. In most cases the specifics relate to the homogenization procedure. The purification steps are the same for all tissues.

Unlike the liver mitochondria, the respiratory activities of mitochondria from heart, brain or kidney do not depend on the metabolic state of the organism (fed or starved). However, respiratory activities depend on phenotypic metabolic properties of animals and the mitochondria. Oxidation of some substrates, for example succinate (no rotenone present), can be inhibited due to intrinsic inhibition of SDH by oxaloacetate, which in heart mitochondria of the Sprague Dawley rats does not depend on whether RHM were isolated with or without BSA.

Another problem, which a mitochondriologist faces when studying mitochondria of a particular organ, is that mitochondria from different parts of the organ or cell types, have differences in structure and functions. For the liver these are differences between mitochondria from periportal and pericentral hepatocytes. For the heart and skeletal muscle, these are differences between mitochondria of the subsarcolemmal and interfibrillar origin. In brain, different parts of the brain and spinal cord may have differences in mitochondrial metabolism. These problems are very poorly studied biochemically. However, we have to keep in mind that during isolation of the liver mitochondria we have a mixture of periportal and pericentral mitochondria that functionally are similar but also have some differences, which we usually do not take into consideration. When isolating heart and, particularly, skeletal muscle mitochondria, it is very important whether we use protease (Nagarse) for the tissue digestion, or not. Without treatment with Nagarse, the skeletal muscle tissue will yield subsarcolemmal mitochondria and the yield will be very small, even if the tissue was disintegrated with a Polytron grinder. Without Nagarse, the Polytron grinded heart will also yield predominantly subsarcolemmal mitochondria, but the yield will be much higher than for the similar amount of skeletal muscle tissue. Nagarse dramatically increases the yields of mitochondria, particularly for the skeletal muscle. However, I have evidence that

Nagarse may change some properties of the isolated mitochondria. Though, this issue has to be studied more thoroughly.

Homogenization

It is useless to homogenize the muscle tissue with the glass/Teflon homogenizer. Before homogenization, the tissue must be minced, that is cut into small pieces as much as possible. The usage of Thomas slicer is good enough for mincing, but it is messy and time consuming. Any kind of Polytron-type homogenizers (such as BioSpec shown on Fig. 3.4) is much better than the Thomas slicer. The Polytron-type homogenizators can be used with other hard tissues such as kidney, and even with the liver for maximizing the yields. In general, wide Polytron tubes are more preferable because you may shorten activation time. With kidney and liver usage of proteases is unnecessary. Even with the heart tissue the yields are high enough without Nagarse. However, there is a problem regarding the origin the mitochondria. Without Nagarse the mitochondria might be predominantly subsarcolemmal. With Nagarse the yields are much higher because of the interfibrillar mitochondria.

With skeletal muscle tissue the yields without Nagarse are very low. Therefore it is necessary to use protease treatment of the minced tissue. However, the properties of the interfibrillar mitochondria in the skeletal muscle are different from those of subsarcolemmal mitochondria. In part, this might be associated with the effects of the protease. Therefore, you should minimize the exposure of the tissue to protease by liming time, and wash the digested tissue with buffer 2-3 times before homogenization with Polytron.

The tip of the chilled Polytron tube is placed at the very bottom of the 25-50 ml centrifugal tube containing the minced tissue and 10-25 ml of the isolation buffer, and turn on the motor just for 2-3 seconds, once or twice. With skeletal muscle the procedure can be repeated 2-3 times. Allow the pieces of the tissue to settle down to the bottom before the next activation of the grinder. It is important to avoid formation of air bubble and foam. This will increase oxidative damages to mitochondria. After that, dilute the homogenate 2-3 times with the isolation buffer and homogenize with the conventional glass/Teflon homogenizer at 300-400 rpm, just 2-3 strokes up and down. The kidneys and hearts of mice are small; therefore homogenization of the heart and kidneys from a mouse is better to perform in 25 ml conical tubes.

Table 3.2. Composition of the isolation buffer.

Components	Mol. Weight	1 Liter
75 mM Mannitol	182.2	13.67 g
175 mM Sucrose	342.3	59.9 g
10 mM MOPS, pH 7.2	209.3	2.093 g
1 mM EGTA (Ca-free)#	380.4	380.4 mg
0.1% BSA Fraction V, Σ A4503-50G	≈ 66,000	1 g

IMPORTANT. The presence of 0.1% BSA basically is not necessary for the heart mitochondria because it does not affect the degree of the intrinsic inhibition of succinate dehydrogenase by oxaloacetate. Rat heart (RHM) and mouse heart (MHM) mitochondria show excellent rates of respiration and respiratory control ratios even if isolated without BSA. However, with succinate as a substrate, the intrinsic inhibition of SDH will persist even in the heart mitochondria isolated with BSA. Of more importance is how long the heart's tissue was exposed to Nagarse. That is why it is better only to perfuse the heart with Nagarse and then cut the heart into pieces in a medium without the enzyme and rinse 2-3 times the heart's pieces with a large (10-15 ml) volume of the isolation buffer. Nagarse for perfusion and exposure of the minced tissue (skeletal muscle) is used at 3 mg Protease in 10 ml of the isolation buffer WITHOUT BSA.

IMPORTANT. The heart and skeletal muscle mitochondria are very large and branched. Therefore physical disintegration of the tissue will yield large amount of physically damaged mitochondria. Damaged mitochondria will have high ATPase activity, decreased State 3 respiration and increased state 4_1 (that is after phosphorylation of added ADP) due to the ATPase activity, and thus complicate experimental results in some other ways. Therefore, it is better to eliminate the damaged mitochondria. This can be achieved by purification of the mitochondria using the Percoll gradients. After preparation of the final working suspension of the isolated mitochondria, let the mitochondria rest on ice for about 30-40 minutes. This will allow some damaged mitochondria to seal the broken membranes and "recover'. I noticed that this simple procedure makes experimental measurements much more reproducible. If you begin to do experiments within 20-30 minutes past isolation, often the first 1-2 experiments differ from the later ones. While mitochondria are resting, you can determine the protein concentration, prepare instruments for the experiment and eat lunch. When stored on ice, the mitochondria preserve their quality for 4-6 hours. I usually repeat the first two experiments at the end, after I finished the planned experiments (of course, if there were enough mitochondria).

Purification of mitochondria using the discontinuous Percoll gradient

Table 3.3. Preparation of the Percoll discontinuous gradient.

% of Percoll V/V	Percoll (ml)	Isolation Medium (ml)
40%	40 ml	60 ml
23%	23	77
15%	15	85

IMPORTANT. Percoll is used for the isolation of cells, organelles, and viruses by density centrifugation. Percoll consists of colloidal silica particles of 15-30 nm diameter (23% w/w in water), which have been coated with polyvinylpyrrolidone (PVP). Percoll is very sensitive to physical damage, and therefore should be mixed with the isolation medium by gentle shaking rather than by pipetting in and out. The solutions should be stored in a cold room for no longer than 3 - 4 days. Therefore, prepare no more than 50 – 100 ml of each Percoll solution for a week. Do not store Percoll in refrigerator for several years because it will deteriorate and the gradients will be bad.

Percoll has a rather alkaline pH, however, do not bother with this and prepare the solutions using regular isolation buffer. I made attempts to adjust pH of the Percoll solutions to pH 7.2 and found that this results in worsening of the Percoll's ability to maintain gradients. Mitochondria withheld well the brief exposure at low temperature to alkaline pH during the purification procedure.

Percoll is well suited for density gradient experiments because it possesses low viscosity compared to alternatives, low osmolarity and no toxicity towards cells and their constituents. I prefer Percoll over Ficoll because Percoll is much easier to use, requires less time and gives reproducible results. Ficoll is a neutral, highly branched, high-mass, hydrophilic polysaccharide which dissolves readily in aqueous solutions. Ficoll radii range from 2-7 nm. It is more commonly used for fractionation of the blood samples. Though, some researchers used it also for purification of mitochondria. However it requires much more time because creating density layers is a much slower process as compared with Percoll.

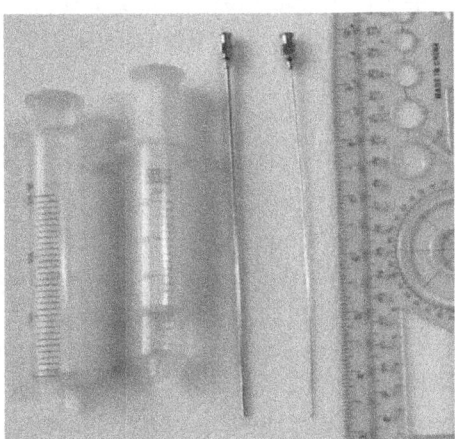

Figure 3.9. Syringes and long surgical needles for creating Percoll gradients.

The thicker needle is used to fill the syringe with sticky viscous Percoll solution. The thin needle is used to feed the Percoll solution to the bottom of the tube. The left syringe is bad for this procedure because the needle has to be fixed tightly by screwing into the syringe (such as on the right) using pliers. There must be no leakage of the buffer at the needle. The needles are about 7 inches long.

The Procedure of Purification of Mitochondria with the discontinuous Percoll gradient

For purification you have to use the Beckman-Coulter swinging-bucket rotors. I describe the procedure for the SW-41 swinging-bucket rotor commonly used for several models of the Beckman-Coulter ultracentrifuges. For the SW-41 rotor I used narrow Nalgene clear tubes (26 x 89 mm for Beckman) about 12 ml volume. These tubes allow easily see the layers. However, after 4-5 spins, they often crack. To avoid the loss of precious experimental mitochondria, replace all tubes after using them 4 times, without waiting for the disaster. The tubes are cheap, but experimental animals are expensive. For example, one transgenic rat, expressing human mutated SOD1 gene, cost $100, plus all other expenses during experiment, and the precious time.

IMPORTANT. Before experiment, place the rack for the SW-41 rotor swinging metal tubes into a plastic box with ice, and insert into the rack the metal tubes: the front row from left to right 1, 2, 3; and the second row 4, 5, 6. Insert the Nalgene tubes. This will cool the tubes in advance. Mark the front side of both the rack and the plastic container by writing "FRONT". This will eliminate even the slightest possibility to misplace the tubes.

Making the Percoll gradient

Step 1. Place in the 12 ml Nalgene centrifugal tubes 3-4 ml of the mitochondria to be purified suspended in 15% Percoll;

Step 2. Take a 40–60 ml syringe; fill it slowly, using the long needle with the larger diameter (see Fig. 3.9), with the 23% Percoll. Change the large needle for the long needle of smaller diameter, put the tip of the needle to the bottom of the tube, and very slowly inject 4-5 ml of the 23% Percoll. Remove the needle slowly.

Commentary. Use small pliers to tighten the needles to the syringe. This will exclude leakage.

Step 3. Fill the syringe with the 40% Percoll, and very slowly without disturbing the layers repeat the procedure described above by injecting 2-3 ml of 40% Percoll.

Commentary. If you have small amount of tissue, use only two or four opposite tubes: 1 & 4; 2 & 5. The unused metal tubes must be tightly closed without inserted Nalgene tubes. Otherwise they will become smashed.

Step 4. Tighten the lids and carefully carry the box with the tubes to the Beckman centrifuge without shaking and tilting.

Commentary. If you were careful in adding precisely the amounts of the gradients, you can avoid weighing the tubes for balancing. This will save you time.

Step 5. The centrifuge should be turned on and chilled in advance with the SW-41 rotor inside or store the rotor in the cold room.

Step 6. Handle the tubes and the rotor with installed tubes carefully. Transfer the tubes to the SW-41 rotor and install the rotor into the Beckmann centrifuge. The rotor should slightly strike the bottom to ensure that it was set down fully.

Step 7. Make the following settings: Temperature: 4°C; Time: 15 min; Speed: 31,000 g; set Acceleration & Deceleration at LOW Spin.

After the centrifugation was completed, carefully carry the tubes back to the cold room and open the leads. Using Pasteur pipette and vacuum suction remove the layers above the layer of mitochondria just above the bottom 40% Percoll. When the tip of the pipetter gets close to mitochondria, press the tip to the tube's wall slightly above the surface of the liquid, so it will suck in the solution slowly. In this way you can remove as much of the liquid above mitochondria as possible. But be careful! Better live some of the liquid, rather than suck in mitochondria.

Thus, for a 12 ml tube for the Beckman swing rotor SW-41 use 3-4 ml of 15% mitochondrial solution, 4-5 ml of 23% Percoll, and 3-2 ml of 40% Percoll. If you must use other type of rotor, change the volumes keeping in mind that the middle volume containing 23% Percoll solution must be long enough (by height) to allow separation. You may vary the 15% Percoll, or cut down 40% Percoll to 2 ml, but keep the 23% layer volume as large as possible.

The mitochondria can be removed by usual plastic transfer pipettes. To be sure that you collected all mitochondria, you can suck in also some amount of 40% Percoll. Place mitochondria in the tube for SS34 rotor containing 20-25 ml of the isolation buffer.

Step 8. To remove the remnants of Percoll from collected mitochondria, re-suspend mitochondria in 30 ml of the isolation buffer. Centrifuge (Sorvall) at 16,000 g for 10 minutes. Higher speed is necessary to remove Percoll.

Step 9. Remove the supernatant by suction. Be careful, the sediment may be loose because of remaining Percoll. Resuspend mitochondria in 2-3 ml of isolation buffer and transfer into a new tube containing 20-30 ml of isolation buffer, and spin at 9000 g for 10 minutes.

Step 10. Collect mitochondria and make working mitochondrial suspension using "Full" incubation buffer.

Purification of mitochondria using Continuous Sucrose-Percoll gradient

In some cases, particularly if you do not have expensive Beckman-Coulter refrigerating centrifuge, or a swing bucket rotor, you can use Sorvall refrigerating centrifuge and SS34 rotor. In this case for purification of mitochondria use the continuous Sucrose-Percoll gradient.

Table 3.4. Preparation of the 1.25 M Sucrose.

Components	Mol. Weight	100 mL
1.25 M Sucrose	342.3	42g 790 mg
10 mM MOPS	209.3	209.4 mg
1 mM EDTA	292.2	29.2 mg

The Continuous Sucrose Percoll gradient is prepared by spinning down the mixed components with the initial (unspinned) density of 1.06: For this take 7.65 ml Percoll, 7 ml 1.25 M sucrose, 50 µl of 700 mM EGTA, add WATER to the final volume of 35 ml, and thoroughly mix by gentle shaking and turning over the tube. Spin the mixture at 12,000 rpm in Sorvall SS34 rotor for 90 minutes. Store the tubes with created gradient on ice before use (you prepare the gradient just before you start the isolation procedure). At the purification step, layer carefully the mitochondrial suspension (in the isolation medium) on the top of the gradient and spin at 12,000 rpm for 15 min. Remove by suction the upper part of the gradient. Mitochondria are closer to the bottom. At the very bottom will be some Percoll. Remove mitochondria by pipette carefully.

Procedures for isolation of heart mitochondria

Commentary 1. This basic procedure can be used for other types of tissue: kidney, skeletal muscle, brain and spinal cord, and even liver mitochondria.

Commentary 2. In details, preparation of the discontinuous Percoll gradient, purification procedure using Percoll gradient are described above.

Step 1. When removing the heart from a rat, cut off the heart with as long aorta attached as possible. Remove the blood by gently squeezing the heart on a piece of a filter paper and place the heart into a beaker with the ice-slurry cold isolation buffer (mixture of the liquid and frozen medium). The heart from a mouse is simply cut off, placed on a filter paper, cut in half and then placed in a beaker with the ice-slurry cold isolation buffer.

Step 2. Using two small pincers hold the sides of aorta and hang the rat heart by inserting the needle of a sufficiently large diameter. The sharp end of the injection needle must be cut off and the rims smoothed. Fix the heart to the needle by small surgical forceps and then by tying aorta to the needle with a strong surgical thread. Do not remove the forceps to prevent sliding off the heart during perfusion. Perfuse slowly the rat heart with 20 ml of epy isolation buffer (top syringe) and then with 10 ml of the isolation buffer **without BSA** containing 0.3 mg/ml Nagarse [Sigma protease Type XXVII]. It is convenient to use a two way connector to which the needle and the syringes are attached.

NOTE: The isolation buffer with Nagarse MUST not contain BSA.

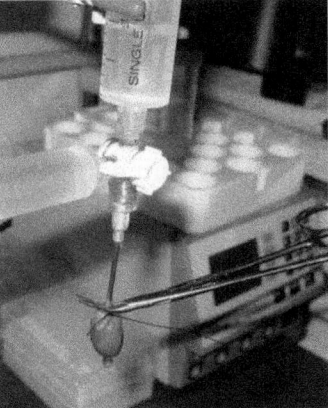

Figure 3.10. Fixation of the rat heart on a surgical needle for perfusion.

Step 3. Cut down the heart from the needle and trim off atrium, fat tissue, cut open on a filter paper (to remove medium with Nagarse), and place into small beaker containing 5 ml of the ice-cold isolation buffer. All procedures are performed at 4° C. Mince the heart with scissors, add more buffer and remove the liquid by filtering through a mesh or cheesecloth. The small hearts from mice are simply exposed to Nagarse solution during mincing procedure and then washed several times with the isolation buffer.

Step 4. Transfer to a 25-50 ml centrifuge tube and add 10 - 30 ml of isolation buffer (depending on the amount of tissue). For the rat heart I used a 50 ml centrifugal tube with conical bottom and a round hole in the tube's cap of the same D as the Polytron's tip. This prevents spilling. For small mice hearts used the 25 ml tube.

Step 5. Homogenize with the Polytron tissue homogenizer (Brinkman PT-10 at 8000) 3 times for 2-3 sec with 10-20 sec intervals between activations to let the tissue peaces to sink down to the bottom. Keep the tip with the knife close to the bottom. Polytron's tube should be cooled on ice before usage and then placed into the tube with the isolation buffer before homogenization. In order to avoid water and ice go inside the Polytron's tube, I cutted off the top of the plastic 50 ml pipetter for cell culturing, placed the Polytron's tube inside and inserted into ice. After finishing disintegration, rinse the Polytron's tip with the isolation buffer by spraying medium through a syringe, and then wash by activating the Polytron in a tube with water.

Commentary: For the heart and particularly for the skeletal muscle tissue the diameter of the Polytron tube must be sufficiently large for rapid and efficient homogenization.

Step 6. Bring the final volume to 30 ml with the isolation buffer, transfer to the glass/Teflon homogenizer and make 3-4 strokes up and down.

Step7. Transfer homogenate to the SS34 tubes and spin at 900 g for 5 min.

Commentary. The heart, skeletal muscle and kidney mitochondria are large; therefore you may lose a lot of mitochondria by spinning at higher "g" during the first centrifugation.

Step 8. Decant the supernatant; discard debris (unless a wash step is included as for the brain and spinal cord mitochondria). Centrifuge the supernatant at 9000 g for 10 min, discard the 2d supernatant.

Step 9. Re-suspend mitochondria in the necessary amount (ml) of the 15% Percoll solution sufficient to fill even number of SW-41 Nalgene tubes using 3-4 ml per tube (see above the description of the Percoll gradient). Mix mitochondria and 15% Percoll by gentle suction in/out through the 1 ml pipette tip.

For purification of mitochondria in the Percoll gradient I used the Beckman swinging-bucket SW-41 rotor. Create the discontinuous Percoll gradient as described above:

Place in the centrifugal tube 3-4 ml of the mitochondria suspended in 15% Percoll;

Fill a 30 ml syringe, using the long large needle, with the 23% Percoll. Change the large needle with the long needle of smaller diameter, put the tip of the needle to the bottom of the tube, and very slowly inject 4-5 ml of the 23% Percoll. Remove the needle slowly.

Fill the syringe with the 40% Percoll, and inject 2-3 ml to the bottom of the tube.

Step 10. Handle the tubes and the rotor with installed tubes carefully. Hang the tubes on the SW-41 rotor and install the rotor into the Beckmann centrifuge. Make the following settings: Temperature: 4oC; Time: 15 min; Speed: 31,000 g; set Acceleration & Deceleration at LOW. Spin.

Step 11. Remove, using the vacuum suction and the Pasteur pipette, the layers above the mitochondria. The mitochondria are located just above the bottom layer of 40% Percoll. Using plastic pipette carefully collect mitochondria and transfer them into a centrifuge tube containing 20 ml of the isolation buffer. Erythrocytes should remain at the bottom, below the 40% Percoll.

Step 12. Re-suspend mitochondria in the isolation buffer. Centrifuge (Sorvall) at 16,000 g for 10 minutes. Higher speed is necessary to remove Percoll.

Step 13. Remove the supernatant by suction. Be careful, the sediment may be loose because of remaining Percoll. Resuspend mitochondria in 2-3 ml of isolation buffer and transfer into a new tube containing 20-30 ml of isolation buffer, and spin at 9000 g for 10 minutes.

Step 14. Collect mitochondria and make working mitochondrial suspension using "Full" incubation buffer.

Isolation of skeletal muscle mitochondria
Preliminary steps.

Prepare Percoll (15%, 23% and 40%) solutions as described above. Keep on ice before use.

Prepare the ice-slurry cold isolation medium for collecting muscle tissue by mixing liquid and frozen isolation buffer, which must constitute at least 40% of the total volume.

Put all tubes, glass and Polytron homogenizers, beakers with the media on ice.

Step 1. Sacrifice the animal, and after that WORK IN A COLD ROOM. Collect muscles from the animal's hindquarters (hind legs with the lower section of the spine muscles). Start stripping muscle tissue from the paws by cutting tendons along the bones. Working with scissors along the bones you can get large chunks of meat. Try to avoid little pieces because they will be not sliced well in the Thomas slicer. If you will use Polytron, the small pieces are even desirable. Put the muscle tissue into a beaker with the ice-slurry isolation medium kept on ice.

Step 2. Place the muscle tissue into a large Petri dish, dry from the medium. Remove fat and connective tissue from the samples, working on a Petri dish placed on ice. Weigh the tissue.

Step 3. Tissue processing. There are several options. 1) Use the Thomas slicer; or, mince the tissue with scissors. 2) Place the minced tissue into a beaker containing 2 ml/g tissue of the isolation buffer without BSA and containing 0.3 mg/ml Nagarse. For the tissue around 8-12 g use 20 ml of the Nagarse solution. Incubate for 1-2 min while continuing mincing the tissue further with scissors.

Because Nagarse can digest mitochondrial membrane's proteins (for example ANT), try to shorten the exposure of the tissue to the protease. For this, remove the liquid containing Nagarse from the beaker using cheesecloth, and wash the tissue 2 times with 20 ml of the isolation buffer before treatment with Polytron or the Glass/Teflon homogenization. Blot the rest of the fluid with Kimwipes tissue or filter paper.

IMPORTANT. With the muscle tissue not treated with Nagarse, any homogenization (Thomas slicer + homogenization with the glass/Teflon pestle, or with Polytron) will give mostly subsarcolemmal mitochondria with the very low yields. With Nagarse, you will get also subsarcolemmal mitochondria with much larger yield of the interfibrillar mitochondria. If you avoid over-digestion (good washing, shot exposure to the enzyme) and follow instructions, the quality of mitochondria will be good.

Step 4. Homogenization. It is preferable to use the Polytron-like tissue homogenizers followed by 3-5 strokes in the glass/Teflon homogenizer, as described for the heart tissue.

Commentary. For the heart, and particularly for the skeletal muscle tissue, the diameter of the Polytron tube must be sufficiently large for rapid and efficient homogenization. With the narrow tube, the homogenate will repeatedly pass through knives, and this will damage mitochondria.

The following steps are precisely the same as described above for the isolation of heart mitochondria.

Procedures for isolation of kidney mitochondria

Preparation of the kidney tissue.

When you work with a rat, remove kidney(s), cut away fat, vessels, and cut the kidney in half along the long frontal axis using sharp knife or scalpel. Remove the middle medullar part of the kidney. Cut the remaining tissue into small pieces with scissors and rinse two times with the isolation buffer. If you work with mice, remove kidneys, clean from fat and chill kidneys in a beaker containing the ice-slurry isolation medium.

Continue further procedures from the Step 5 (Polytron homogenization) as described for the heart mitochondria.

Commentary #1. Rat kidney tissue is tough and difficult to homogenize properly using the glass/Teflon homogenizer or a Dounce homogenizer. You will get something, but the quality of mitochondria will be not good and reproducibility will be low. If you do not have Beckman centrifuge, use the continuous Percoll gradient for Sorvall centrifuge as described above.

I have isolated kidney mitochondria from rats and mice following exactly the procedure for the heart mitochondria and purifying with Percoll gradient. The mitochondria showed very high respiratory rates and respiratory control ratios. The yields of mitochondria from two small kidneys of one mouse are usually higher than from the much larger rat kidney.

Commentary #2. With kidney mitochondria after purification with Percoll gradient you may find that a large portion of mitochondria go INSIDE the 40% Percoll. Collect this fraction. These are good mitochondria. Because of a large size they go into 40% Percoll. Some erythrocytes may still be present, but they will disappear during the 2 steps of the final washing.

Isolation of brain and spinal cord mitochondria.

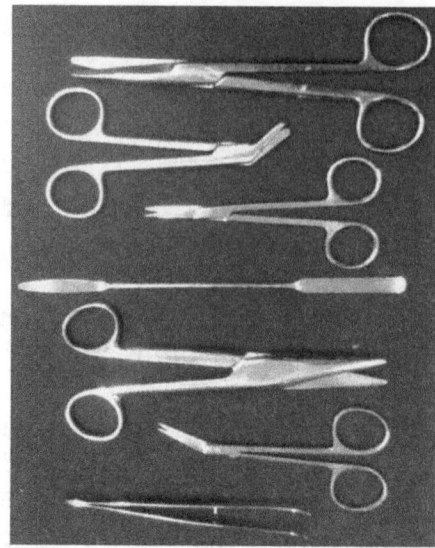

Figure 3.11. Surgical instruments for extracting brain and spinal cord.

From top down:

• Scissors for cutting off mouse's head and opening the rat's abdomen;

• Scissors for cutting rat's skull (wire cutter);

• Small scissors for cutting mouse's skull and mince tissues;

• Spatula with a slightly bend tip for extracting brain from the skulls;

• Large scissors to cut across the rat's spinal cord column;

• Curved scissors for cutting trough the rat's spinal column;

• Tweezers for manipulating and cleaning the tissue.

Procedures for isolation of brain mitochondria

Step 1. A rat, or a mouse, is quickly killed by decapitation after general anesthesia. Rapidly extract the brain. With the mouse's skull, cut using small scissors through the foramen magnum first horizontally on both sides, and then vertically along the sagitall suture. The sides of the cranium are separated with a pincer, and the brain removed using slightly curved narrow spatula. The bones of the rat's skull are much thicker. I found very convenient to use the surgical scissors for cutting wires (see illustration). I usually removed cerebellum and the brain stem. The remaining brain is called the forebrain, which I used for isolation of mitochondria. According to David Nicholls (personal communication), mitochondria from cerebellum are very distinct from those of the forebrain. Place the forebrain into the ice-slurry (mixture of liquid and frozen buffer) isolation buffer.

Step 2. In a cold room remove with fine tweezers the film with blood vessels that cover the brain (but do not be too picky about that). If you use 40% Percoll, you can skip this procedure altogether. Mince the brain finely using scissors in a small amount (10–20 ml) of the isolation buffer. Transfer to the Dounce homogenizer (40 ml volume), and add 30-35 ml of the isolation buffer.

Step 3. Hand homogenize the tissue using a Dounce homogenizer. For 2-3 brains you can use 2 tubes for Sorvall SS34 filled (one finger to the top) with the isolation buffer. Use one tube for homogenization of up to 3 brains. If you isolate mitochondria from 4 brains, than it is better to homogenize 2 brains separately and use 4 SS34 tubes. Perform 20-25 up/down sharp strokes using **loose-fitting pestle** (clearance 0.12 mm) only. For this, you push the pestle down slowly to the bottom and then sharply pull it up, covering the outlet with another hand around the stem of the pestle. You have to wear rubber gloves. This will prevent spills and limit the lift of the pestle. The idea is to create a kind of low pressure that causes the synapses to blow up. I found that the usage of the tight pestle deteriorates mitochondria significantly. After the brain was homogenized, add medium from the second tube into Dounce and mix with 1-2 passages of the pestle.

Step 4. Centrifuge the homogenate at 1000 g for 5 min.

Step 5. After centrifugation, decant the supernatant into a new tube but without getting the "tail" of the white staff from the centrifugal tube. Keep the supernatant on ice. Add 20 ml of cold buffer to the tube with the sediment and re-suspend the pellet using the ball of the loose pestle. Repeat the procedure with the second tube containing the pellet after the 1st spin.

Step 6. Repeat Step 4, decant the supernatants and discard the pellet.

Step 7. Combine the two supernatants from steps 4 and 6 and centrifuge at 9500 g for 10 min.

NOTICE. Re-suspension of the pellet after the 1st centrifugation in the additional amount of isolation buffer (actually washing), will increase the yield of mitochondria by about 20-30%. Particularly, if the initial homogenate was thick (a homogenate of 3 rat forebrains).

Step 8. Re-suspend mitochondria in the necessary amount (ml) of the 15% Percoll solution sufficient to fill even number of SW-41 Nalgene tubes using 2.5-4.0 ml per tube (see above the description of the Percoll gradient for more details). Mix mitochondria and 15% Percoll by gentle suction in/out through the 1 ml pipette tip. When I had to simultaneously isolate mitochondria from the brains and spinal cords of the same animals, I usually used 4 tubes for the brain, and two tubes for the spinal cord. Be careful to pare the opposite tubes: #1 versus #4, #2 versus # 5 (for the brain), and #3 versus #6 (for the spinal cord). If you carefully control the volumes you add to the tubes, you can skip balancing of the tubes on the scales.

For purification of mitochondria in the Percoll gradient I used the Beckman swinging-bucket SW-41 rotor. Create the discontinuous Percoll gradient as described earlier. In brief:

Place in the centrifugal tube 2.5-4.0 ml of the mitochondria suspended in 15% Percoll;

Fill a 40 – 60 ml syringe, using the long large needle, with the 23% Percoll. Change the large needle with the long needle of smaller diameter, put the tip of the needle to the bottom of the tube, and very slowly inject 4-5 ml of the 23% Percoll. Remove the needle slowly.

Fill the syringe with the 40% Percoll, and inject 3-2 ml to the bottom of the tube.

Step 9. Handle the tubes and the rotor with installed tubes carefully. Transfer the tubes to the SW-41 rotor and install the rotor into the Beckmann centrifuge. Make the following settings: Temperature: 4ºC; Time: 15 min; Speed: 31,000 g; set Acceleration & Deceleration at LOW. Spin.

Step 10. NOTICE. Unlike the heart or muscle mitochondria purification procedure, with the brain after centrifugation of the Percoll gradient, you will receive a thick top layer of white staff, which is mostly myelin. Remove myelin and the lower turbid layer (synaptosomes) above the mitochondria using the vacuum suction and Pasteur pipette. The mitochondria are located just above the bottom layer of 40% Percoll. Using plastic pipette carefully collect mitochondria and transfer them into a centrifuge tube containing 20-30 ml of the isolation buffer. Erythrocytes will be at the bottom, below the 40% Percoll.

Step 11. Re-suspend mitochondria in the isolation buffer. Centrifuge (Sorvall) at 16,000 g for 10 minutes. Higher speed is necessary to remove Percoll.

Step 12. Remove the supernatant by suction. Be careful, the sediment may be loose because of the remaining Percoll. Re-suspend mitochondria in 2-3 ml of the isolation buffer and transfer into a new tube containing 20-30 ml of the isolation buffer, and spin at 9000 g for 10 minutes.

Step 13. Collect mitochondria and make working mitochondrial suspension using "Full" incubation buffer. Brain and spinal cord mitochondria sediment very rapidly in the sucrose medium. The mitochondrial suspension is much more stable in the incubation medium.

NOTICE. The method I have described is a modification of the method of Sims for isolation of brain mitochondria (Sims, 1990). Originally, according to Sims, the BSA should be added at step 11. Sims recommended adding at this point the fatty acid free bovine serum albumin (10 mg/ml). I have found that you can safely isolate brain mitochondria without BSA, but in some species BM would not oxidize succinate (see my papers Panov et al. 2009, 2010; 2011). However, BM and SCM from diseased animals may be damaged oxidatively during the isolation procedure without BSA. Therefore, I recommend isolating mitochondria from diseased animals using the isolation buffer with 0.1% BSA. If you plan to study the intrinsic inhibition of SDH by oxaloacetate, which is the subject for diversity among species, than do not use BSA. You can release the inhibition of succinate (no rotenone) metabolically (Panov et al. 2009, 2011).

Procedures for isolation of spinal cord mitochondria

In general, the spinal cord mitochondria (SCM) are isolated by the same procedure as the brain mitochondria. Therefore, I describe here only the technique of deriving the spinal cord tissue. The most important items during this procedure are time and temperature. From the moment of the spinal cord tissue homogenization in a Dounce homogenizer, the isolation procedures are the same as for the brain mitochondria. The necessary instrumentation is shown on Fig. 3.11.

Getting the spinal cord tissue.

Step 1. After the anesthetized (rat) animal was decapitated, open the abdomen and chest from the bottom through diaphragm and chest to the neck.

Step 2. Take a gauze tissue napkin and eviscerate all organs from the chest and abdomen down.

Step 3. With strong scissors cut through the spine just above the hind legs, and cut out the spine removing as much muscle and fat tissue as possible but not spending too much time on this procedure.

Step 4. Place the removed spine column into a large volume (about 300-500 ml) of the ice-slurry cold isolation buffer containing at least 40% of the frozen buffer. The procedure of removal of the spinal column, from the time of decapitation to placement into ice-slurry medium, usually takes about 1 minute.

Proceed with other animals. By cutting away as much muscle tissue and fat from the spine you shorten the time for chilling.

Step 5. In the Cold Room use fine but strong scissors to cut through the spine by inserting one branch of the scissors into the spine cord channel and cutting close to the vertebrae. Cut on both sides from the vertebrae.

Step 6. Collect the spine cord tissue with fine pincers, and place the tissue into a small beaker with chilled isolation medium but without ice.

Step 7. Remove carefully using fine pincers the films, strips of muscles and blood vessels from the tissue, and cut finely with scissors in a small volume of the isolation buffer. When the connective tissue and muscle are removed, the spinal cord tissue becomes very soft.

Step 8. From this step, homogenization in the Dounce homogenizer, follow the procedures for Step 3 described for the brain tissue. But the first up-down movements of the loose fitting pestle are more difficult than with brain tissue. It depends how carefully you removed the muscles and connective tissue. You can after 3-4 strokes wipe out the brim of the pestle's glass ball where connective tissue has accumulated. After that the strokes will be easier.

Follow further from the **Step 4** the procedures for isolation of the brain mitochondria.

Isolation of mitochondria from cultured cells

IMPORTANT. To study mitochondrial functions in whole cells or isolated cell mitochondria, the cells MUST be cultured at all times in the absence of antibiotics, particularly aminoglycoside antibiotics - gentamicin and streptomycin. These antibiotics by a number of mechanisms, described in the literature, make mitochondria dysfunctional.

I have isolated mitochondria from several types of cultured cells (prostate cancer cells PC-3 and DU145, rat hepatocytes, human lymphoblastoid cells) grown in the presence of streptomycin and in all cases the isolated mitochondria had zero respiration on any substrate.

Cells for isolation of mitochondria have to be harvested at the density of 60 -70% confluence. Some fast growing cells, such as prostate cancer cells even at 70% confluence show regions, which are actually 100% confluent for that area. Therefore, fast growing cells have to be harvested at 50% confluence. The reason for this is that in confluent cells some mechanisms trigger apoptosis, and therefore isolated mitochondria will have much lower membrane potential.

Example. In isolated mitochondria from the prostate cancer cell lines PC-3, DU145 and LNCaP I have estimated the membrane potential close or higher than 200 mV. In mitochondria from the same cells harvested at 80–90% confluence the membrane potential was only 160 mV. The membrane potential was measured using the TPP-sensitive electrode (Panov & Orynbayeva, 2013).

Isolation of mitochondria from cultured cells using Digitonine.

This is the procedure used by many researchers and with this method I begun to isolate mitochondria from the cells. The method uses digitonin to disrupt the cells. The idea was that digitonin affects only membranes rich in cholesterol, such as the cytoplasmic membranes. Because mitochondrial membranes are believed to have very little cholesterol, digitonin should not damage mitochondria. This method worked fine with cultured human lymphoblastoid cells. However, it did not worked at all with the prostate cancer cells. These cells were destroyed only by 50-60% even in the presence of 4-fold higher concentration of digitonin. Therefore, I have developed my own alternative method, which does not require digitonin. I named this method as the "Hypotonic Swelling Method", which I and several of my colleagues used successfully for isolation of mitochondria from various types of cells.

The "Classical" method of isolation of cell mitochondria with digitonin.

Step 1. Harvest cells in mid-log growth (approximately $2\text{-}3 \times 10^9$ cells, about 1.5 liters fibroblasts or 3 liters spinner bottle cultures, e.g. 143B derived cybrids) by pelleting in 250 ml bottles, 300 g (1300 rpm Jouan centrifuge) x 3 min.

(10 flasks half-confluent LNCaP cells give about 2.5 g of cells, 3 big rippled bottles of PC-3 cells - 2 g, DU145 - 2.1 g).

Some cells have to be scraped or tripsinized. Try to use as little tripsin as possible. I used three times less than recommended tripsin, incubated for 2 minutes at 37oC, and then used a rubber mallet to detach the cells with a sharp blow. Sometimes I didn't even have to use tripsin to detach the cells.

Step 2. Transfer pellets to a pre-weighed 50 ml tube, re-suspend in the cold isolation buffer. Centrifuge again at 650 g x 3 min. Remove the supernatant and note weight of the cell's pellet. Thoroughly re-suspend cells in 4 ml buffer for each gram of packed cells (Final volume x 5, i.e. 5 times larger than the initial volume of cells).

Step 3. While swirling the cell's suspension in a tube, add digitonin (10% w/v in DMSO) slowly to final concentration of 0.1 mg/ml [this is 1 µl of 10% Digitonin/1 ml of suspension]. Gently mix for 1 min with pipette and check an aliquot of cells with Trypan blue. Add more digitonin in 0.05 mg/ml aliquots until >90% of the cells are permeabilized. The optimal final digitonin concentration to achieve this varies between 0.2 and 0.45 mg/ml. You have to be quick, as the excessive digitonin can drastically reduce yield and quality of mitochondria.

Step 4. Increase the volume of the suspension 10 fold with the isolation buffer and centrifuge at 3000 g for 5 min. This will remove the excess of digitonin.

Step 5. Remove the supernatant with pipette (do not pour it as the slimy digitonin-treated cell pellet will slide out too) and re-suspend in the total volume of x 5 with the isolation buffer, transfer to chilled Dounce homogenizer and disrupt cells with 10 to 20 passes using a tight-fitting pestle. Check microscopically for adequate cell disruption.

Step 6. Triple the homogenate's volume with the buffer and centrifuge at 625 g for 3 min. Repeat spins with the supernatant two more times, or until very little material (unbroken cells, nuclei) remain in the sediments.

Step 7. Centrifuge supernatant at 12,000 g for 10 min using SS-34 rotor (Sorvall). Use 30 ml Nalgene polysulfonate tubes. Pour off the supernatant, allowing most of the light sediment ("halo") to detach from the buff (brownish) mitochondria pellet. Gently rinse the pellet with few milliliters of cold buffer to remove the remaining light material before re-suspending the mitochondrial pellet in 25 ml buffer and centrifuge again at 12,000 g for 10 min.

Step 8. Pour off the supernatant, allow the tube to stand on ice a few minutes, remove the last drops of buffer. Re-suspend mitochondrial pellet with 0.1 ml of buffer per each 1 g of cells used to give a net protein concentration of around 10-15 mg/ml. Measure the final suspension volume to determine the pellet volume, so that correction for the BSA in the isolation buffer can be made when protein is measured.

References:

• Moreadith, R., Fiskum, D.R. (1984) Isolation of mitochondria from ascites tumor cells permeabilized with digitonin. Anal. Biochem. Vol. 137, 360-367

• Nowell, N., Nalty, M.S. (1986) A digitonin-based procedure for the isolation of mitochondrial DNA from mammalian cells. Plasmid. Vol. 16, 77-80

• Trounce, I.A., Kim, Y.L., Jun, A.S., Wallace, D.C. (1996) Assessment of mitochondrial oxidative phosphorylation in patient muscle biopsies, lymphoblasts, and transmitochondrial cell lines. Methods Enzymol. Vol. 264, 484-509

Isolation of mitochondria from cells using the Hypotonic Swelling

Original Method **was** described in: Panov et al. (2005) Mol. Cell. Biochem. 269, 143-152.

Table 3.5. Isolation buffer

Components	Mol. Weight	1 Liter
75 mM Mannitol	182.2	13.67 g
225 mM Sucrose	342.3	77.02 g
10 mM MOPS, pH 7.2	209.3	2.093 g
1 mM EGTA	380.4	380.4 mg
0.1% BSA Fraction V, Sigma A4503-50G	≈ 66,000	1 g

Table 3.6. Hypotonic buiffer.
Preparation of 120 mosM sucrose medium.

Component	Mol. Wt	1 Liter
Sucrose 100 mM	342.3	34g 230 mg
MOPS 10 mM, pH 7.2	209.3	2.1 g
EGTA 1 mM	380.4	380 mg
BSA 0.1%	≈ 66,000	1 g

Table 3.7. Hypertonic 1.25 M sucrose for adjustment of tonicity.

Components	Mol. Weight	100 mL
1.25 M Sucrose	342.3	42g 790 mg
10 mM MOPS, pH 7.2	209.3	209.4 mg

IMPORTANT. This method does not require detergents and can be used for all cell types. However, for every cell type, the method of osmotic swelling has to be adjusted as described in the Step 3. I also preferred to use the isolation buffer with reversed mannitol/sucrose ratio, which is shown the table below. Mannitol is osmotically active and thus not always is good for some procedures.

Filter sterilize all media.

Step 1. Spin down cells harvested from the culture at 980 g for 5 min.

Step 2. Wash cells with the isolation (Sucrose-mannitol, MOPS, 0.1% BSA) buffer and spin in SS-34 rotor in 50 ml centrifuge tubes at 950 g for 5 min. Remove as much supernatant buffer as possible.

Step 3. Re-suspend the pellet of the harvested cells in a hypotonic medium of 120 mOsM tonicity (Table 3.6) approximately 5 ml of the buffer per 1 g of cells. Compare cell's volume under microscope before and after. If not all cells become swollen, take a 1 ml sample of the suspension of your cells in the hypotonic solution and add an aliquot of water, say 100 μl. Look again under microscope. Repeat the procedure with the cells in 1 ml sample until all cells will be swollen. After that add the equivalent amount of water to you main suspension of the cells, and check under microscope.

Step 4. After incubation for 5-7 minutes, transfer cells to chilled Dounce homogenizer and disrupt cells with 20 to 30 sharp strokes using a tight-fitting pestle. For this, slowly move the ball of the pestle to the bottom, and then with a sharp movement move the ball up. This will create a low pressure, which will blow up the swollen cells. Check under microscope for adequate cell disruption.

Step 5. Add the corresponding amount of hypertonic 1.25 M sucrose (see Table 3.7 above for preparation) to bring up tonicity to 320 mOsm (≈1.05 ml of 1.25 M sucrose per each 10 ml of the low tonicity suspension). Mix with a gentle movement of the loose pestle and double or triple the volume with the regular isolation buffer.

Step 6. Spin the disrupted cells on Sorvall SS-34 rotor at 930 g for 5 minutes.

NOTICE. In some cells mitochondria are very large, they have circular stricture and thus are very heavy. Therefore, you may lose much of the mitochondria. Therefore, I recommend trying sediment the cell's debris at lower g. Even if the supernatant will be contaminated with the cell's debris you can later purify mitochondria using the Percoll gradient.

Step 7. Spin the supernatant at 9000 g for 10 minutes/

Step 8. Remove from the pellet all fluffy, light staff living brown sediment, and re-suspend the remaining brown pellet in the isolation medium, and spin for 5 min. at 930 g;

Or alternatively, use the Percoll gradient, as described for the heart and brain mitochondria, to purify mitochondria.

Step 9. Spin the supernatant at 9000 g;

Again, remove all light staff, and re-suspend mitochondria in the "Full" medium without BSA and EGTA.

NOTICE. If you purified your cell mitochondria with the Percoll gradient, follow the procedure described above for the heart mitochondria.

Commentary. The usual yields from Human lymphoblastoid cells vary from 0.3 - 0.6 mg/g.

CHAPTER 4

Preparation of the Working Suspension of the Isolated Mitochondria & Determination of Protein Content

The isolated mitochondria have to be suspended at a reasonably high concentration for storage on ice for future experiments. Most people suspend isolated mitochondria in 0.25 M sucrose solution with the buffer. This method works very well with the liver mitochondria. However, I have found that mitochondria from spinal cord and brain mitochondria very quickly sediment and form clamps. This is because sucrose is a neutral chemical. Therefore, mitochondria suspended in the mixture of sucrose and KCl are much less prone stick together to form lumps.

Table 4.1. Sucrose medium for mitochondrial suspension

Components	Mol. Weight	0.25 L
150 mM Sucrose	342.3	12g 840 mg
75 mM KCl	74.56	1g 400 mg
5 mM MOPS, pH 7.2	209.3	262 mg

Lately, I started to suspend mitochondria in the incubation medium for respiration, which I named "Full medium". The medium is based on KCl, contains magnesium and phosphate. These suspensions are even more stable than those made in the sucrose–KCl buffer shown above. However, the type of the buffer for preparing suspension of mitochondria will depend on your experimental tasks.

Mitochondrial suspension procedure. After the final spinning down at the end of the isolation procedure, the mitochondria form a dense spot at the bottom of the centrifuge tube. The remnants of the isolation buffer, which contain EGTA, have to be thoroughly removed using a Pasteur glass pipette and vacuum suction. Let the tube to stand in ice for a couple of minutes and then remove the rest of the medium, which went down from the tube's walls. Suspend the mitochondria in a small, but sufficient volume of the suspension buffer (the buffer for respiration is the best) using a pipetter. For mitochondria from 1 rat brain I used minimum 200-250 µl of the buffer, 1 heart – 300-500 µl, 1 mouse liver - 1 ml. Transfer mitochondria into a 1.5 ml centrifugal tube and thoroughly homogenize by letting mitochondria go in and out of the pippeter's tip. If a lump of mitochondria stick inside the tip, vortex the tip vigorously while it is inside the 1.5 ml tube. If necessary, use another pipetter (100 µl) with a tip having smaller whole to make fully homogenous suspension of mitochondria, use also vortex for the purpose. Preparation of the final well homogenized mitochondrial suspension and measurements of mitochondrial protein concentration are very important steps because they determine the reproducibility of experiments. For this reason the concentration of mitochondria in suspension should not be too high because if you add mitochondria into experimental chamber or cuvette in a small volume (say 5 µl), the small errors in the volume will be multiplied as errors in protein concentration. The practical concentrations of the mitochondria in the final suspension usually vary between 8 and 20 mg/ml. I never trusted to other people the construction of the protein calibration curve and determination of protein.

Mitochondria rapidly form lumps even in the KCl medium, though not as easily as in the sucrose buffer, but they do go to the bottom of the tube placed on ice. Therefore, every time you take a sample of mitochondria from the tube, vortex the tube with mitochondrial suspension for 3-4 seconds at high speed. It is not enough just to touch the vortex. Even if the next sample of mitochondria you take during protein measurement is just 10-15 seconds later from the previous sample, vortex the tube again. This will diminish errors between the parallel measurements.

Determination of protein concentration in mitochondrial suspension by the Biuret method

Reagents:
1). Biuret Reactant. 1.5 g $CuSO_4$ x 5 H_2O dissolve in 500 ml of water and add 6 g of $NaKC_4H_4O_6$ (Sodium Potassium Tartrate) and 300 ml of 10% NaOH. Adjust the volume up to 1 Liter.

2) Make 2% solution of BSA (20 mg/ml). Remember, the solution should be carefully mixed (it easily forms bubbles) and should stand for at least 1 day before use, since protein molecules bind water slowly. Store frozen at -20°C.

3). Solutions for solubilization (Diluent) of mitochondria and other subcellular particles:

Diluent Solution 1: 3% Deoxycholate solution, add 2-3 tablets of NaOH; or you can use
Diluent Solution 2: 0.4% KOH + 0.05% Triton-X100 (I prefer this diluent).

Table 4.2. Protein Determination:

Reactants	Blank	BSA Standard	Mitochondria
	2 tubes	*2 tubes*	*2 tubes*
3% Deoxycholate	0.5 ml	0.5 ml	0.5 ml
H$_2$O	0.1 ml	No addition	0.1 ml
Buffer *(Medium in which Mitochondria are suspended)*	0.1 ml	0.1 ml	0.1 ml
2% BSA	-	0.1 ml	-
Mitochondria	-	-	0.05 ml
Biuret Reactant	2.0 ml	2.0 ml	2.0 ml

Vortex the mixture, heat in the boiling water bath for 40-45 sec., and cool in a water/ice mixture. Read at 540 nm.
[Mitochondria] = (A$_{mt}$ - A$_{blanc}$ / A$_{st}$ - A$_{blanc}$) x 40.7

This is the classical Biuret method I used for many years. The advantage is that you can prepare the Biuret reactant yourself. It will be cheap and reliable, and you will not depend on any Company.
Reference. Gornall, A. G., A. J. Bardawill, M. M. David. (1949) Determination of serum protein by means of the Biuret reaction. J. Biol. Chem. Vol. 177, 751-766.
The major difficulty in preparation of the Biuret reactant is that it is very sensitive to the quality of all chemicals (see Gorball et al. 1949).

The Bradford method of determination of protein concentration with the Pierce Coomassie Protein Assay reagent.

Pierce Coomassie Protein Assay Reagent Kit contains sufficient reagents to perform 190 (380 assays, if to use 2.5 ml instead of 5 ml of the reagent) tube assays. This is a modification of the Bradford Coomassie dye binding calorimetric method for total protein quantitation. The Kit contains:

Coomassie Protein Assay Reagent, 950 ml, containing the Coomassie dye, methanol, phosphoric acid and solubilizing agents in water. Shake well and bring to room temperature before use.

Albumin standard, 10 x 1 ml ampules, containing bovine serum albumin (BSA) at a concentration of 2 mg/ml in a solution of 0.9% saline and 0.05% sodium azide. Available separately as product #23209.

When stored refrigerated, the kit has a shelf life of 12 months. If any kit reagent shows discoloration or evidence of microbial contamination through it away.

Preparation of working reagents for tube assay.

Table 4.3. Preparation of Diluted BSA Standards.
Working Range 100 - 1,500 µg/ml

Vol. of the BSA	Vol. of Solvent	Final BSA Concentration
300 µl (Stock)	0 µl	2000 µg/ml
375 µl (Stock)	125 µl	1500 µg/ml (A)
325 µl (Stock)	325 l	1000 µg/ml (B)
175 µl of (A)	175 µl	750 µg/ml (C)
325 µl of (B)	325 µl	500 µg/ml (D)
325µl of (D)	325 µl	250 µg/ml (E)
325 µl of (E)	325 µl	125 µg/ml (F)

PS. Relationships between [BSA] and extinction are nonlinear, therefore calibration is necessary.

First, prepare the solution to dissolve the mitochondrial membranes and solubilize all mitochondrial proteins. I usually used the following solution (Diluent solution 2, see above): For the 250 ml volume dissolve 1 g of NaOH and 125 mg Triton X-100. Use this solution (Solvent) to prepare multiple concentrations of the mitochondrial suspension for protein determination and dilute BSA standards.

Isolation medium may contain 1-5 mg BSA/ml, therefore to get in the range of the method it is enough to dilute it 10 fold. Mitochondrial suspensions usually contain total protein between 5 -20 mg/ml, therefore prepare 100, 50, 25 and sometimes a 10 fold dilutions for protein determinations.

Table 4.4.
Preparation of Mitochondrial and Isolation Medium Dilutions

µl Mitochondrial Suspension	µl Solvent	Dilution
10 µl (Mitochondria)	240 µl	25 fold
5 µl (Mitochondria)	245 µl	50 fold
5 µl (Mitochondria	500 µl	100 fold
25 µl (Isol. Medium)	225 µl	10 fold

The Standard Protocol: for the tube version.

Step1. For each sample of mitochondria take 3 1.5 ml tubes, and add: Tube #1 0.5 ml, #2 and #3 0.25 ml of the protein Solvent. Into Tube #1 and #2 add 5 µl and into Tube #3 add 10 µl of the mitochondrial suspension. Vortex mitochondria each time vigorously. These are mitochondrial dilutions 100, 50 and 25 fold.

Step 2. Add into the glass (not plastic) tubes, 2 tubes for each dilution, 50 µl mitochondrial dilutions (2 parallel tubes for each dilution). Add 50 µl of the Solvent to the Blank tube.

Step 3. Add into each tube 2.5 ml Coomassie reagent, which was warmed to room temperature.

Step 4. Vortex tubes and measure extinctions at 595 nm against control (Blank) tube, which is 50 µl of diluent + 2.5 ml Coomassie reagent.

Commentary. If you have more than 1 sample of mitochondria, have correspondingly Blanks for each mitochondrial sample. During measurements, use the corresponding Blank for each sample zeroing instrument for the Blank. This will eliminate the time-dependent changes in extinctions. However, avoid to measure at one time more than 3 samples of mitochondria.

IMPORTANT. During experiment, the correct and reproducible determination of protein concentration is one of the MOST important steps. A small error in protein concentration will result in larger dispersion of results, and thus more experiments will be required to ensure statistical significance. This will result in a significant increase in the total cost of the experiments and animal lives. We should always keep in mind the financial and moral sides of the Experimental research. For this reason, I avoided minimization in the volumes of experimental and calibration samples, which increases the chance for errors. And I would never trust protein measurements to a technician and use the method adapted for microplates.

Steps to minimize the errors during protein determination.

During preparation of the working suspension of mitochondria, the mitochondria must be suspended thoroughly using pipetting with the tips with small apertures (such as tips for 100 µl pipetter), intensive vortexing, and KCl-containing medium. The medium for respiration is the best because it delays formation of lumps and sedimentation of mitochondria.

When you take a 5 or 10 µl sample of mitochondrial suspension, vortex the tube with mitochondria intensively for 2-3 seconds, even if you vortexed it 10-20 seconds ago.

Use the same pipetters every day.

When you take a sample of mitochondria for diluting in the detergent solution (Solvent), check the tip and wipe it without touching the aperture. Otherwise, the capillary effect will remove the solution from the tip. Keep the tubes with the solvent open.

After you added mitochondria to the tube with the Solvent, close the tube and vortex the tube thoroughly.

Move the tube, from which you just took the two 50 µl samples of diluted mitochondria, to another row. In this way you will avoid a possibility that you take the sample twice from the same tube.

Always distribute solutions in the rows tubes the same way; begin with the first row and add from left to right.

Tubes for reaction of the mitochondrial samples with the Coomassie reagent must be the glass tubes.

After you added one sample to the glass tube, change the tip.

After addition of the Coomassie reagent to the protein sample vortex the tube.

If you have two or three samples of mitochondria for protein determination, make correspondingly two or three blank controls tubes.

Before measurements of each sample of mitochondria zero the instrument with the corresponding blank.

Make measurements in the spectrophotometer from the lowest (100 fold dilution) to the highest (25 or 10 fold dilution) concentration.

Each new sample of mitochondria should begin with the zeroing of the instrument using the corresponding blank. In this case changes in the optical densities with time will go both in the blank and the samples. In general, make measurements quick after you added the reagent. For this, turn on the instrument **before** you begin determination of protein concentration.

IMPORTANT. Do the measurements yourself, be attentive and concentrate on the task. Do not use glass or quartz cuvettes because they become stained and very difficult to wash. Instead, use plastic disposable cuvettes.

CHAPTER 5

Submitochondrial Particles

Preparation of submitochondrial particles

Submitochondrial particles (SMP) are used for many purposes. The most common applications of SMP are measurements of activities of the mitochondrial respiratory chain complexes under normal and pathologic conditions. For diagnosis of the mitochondrial diseases associated with mutations in mtDNA this is one of the main sources of information. However, in the literature you can find many methodical and methodological errors. For example, some researchers study the activities of the respiratory complexes in mitochondria frozen and thawed several times. These are not SMP, these are mitochondria with damaged membranes. It is difficult to control permeability of the frozen-thawed mitochondria from experiment to experiment, and there is no convincing evidence that this method provides maximum activity. Many researchers do prepare SMP, but do it under wrong or uncontrollable conditions.

The classical method of obtaining SMP involves sonication of whole mitochondria using special instruments. As a result, sonic radiation disintegrates mitochondrial membranes, which then reseal again forming tiny vesicles. The properties of these vesicles are determined by orientation of the proteins embedded into the membrane, which depends on the presence of Mg^{2+} ions (Lee & Ernster, 1967). In the presence of Mg^{2+}, the vesicles resume the right orientation that is the proteins, which were normally facing the matrix space, in the resealed particles also face inside the vesicles. In the absence of Mg^{2+}, these proteins face outside – these are the inside-out particles. The first case is achieved by the presence in the sonication medium of high $[Mg^{2+}]$, the second – by the presence of 2 mM EDTA. If the concentration of Mg^{2+} in the sonication buffer is low, then the orientation of the SMP's vesicles is random, 50-50%.

The whole idea of SMP for studies of the mitochondrial functions is to expose proteins, which face normally the matrix space, to the extra vesicular environment. In the EDTA-SMP with the inside-out orientation of the inner membrane, chemicals and proteins of the matrix, which are not bound to the membrane, are washed out. For example, liver mitochondria contain large amount of catalase. This makes impossible to study ROS production in whole liver mitochondria. During preparation of the EDTA-SMP from the liver mitochondria, catalase is removed, and you can measure the activity of ROS production during oxidation of substrates (Panov et al., 2005). The method described below produces vesicles, which almost by 100% have the inside-out orientation. This renders high reproducibility of the activities measured with these SMP.

Reference: Lee C-P., Ernster L. (1967) Energy coupling in nonphosphorylating submitochondrial particles. In: Methods in Enzymology, Estabrook, R.W. and Pullman M.E., eds., pp. 543-548, Academic Press, New York – London

1). A sample of fresh or thawed frozen mitochondria dilute with the sonication medium (0.25 M sucrose, 2 mM EDTA, 10 mM MOPS, pH 7.2) to contain about 20-30 mg/ml.

IMPORTANT: For phosphorylating particles, instead of EDTA add to the sonication medium 5 mM $MnCl_2$, 10 mM $MgSO_4$, 1 mM succinate Na, or 15 mM $MgSO_4$ and 1 mM ATP (see Lee & Ernster, 1967 for details).

PS. The nonphosphorylating particles have almost by 100% the inside out orientation of the inner membrane, while most of the phosphorylating particles have right orientation. If the concentration of Mg^{2+} is not controlled, than the orientation of SMP becomes random.

2) The suspension is saturated with N_2 and subjected to sonic oscillation 5 times for 2 seconds. The volume of the mitochondrial suspension is 0.5 ml. The tube is placed in a beaker filled with the mixture of ice + water, then add generously dry KCl on top. This will cool the mixture to -15ºC. Between the impulses, cool the sonicator's tip in this cold mixture and then wipe with a tissue and repeat sonication.

3) The sonicated mitochondria are diluted with the double volume of the sonication medium and centrifuge in Beckman (rotor SW41) at 16,000g for 10 minutes to remove the large remnants of mitochondria.

4) The supernatant can be used for most conventional measurements of the respiratory chain activities. Otherwise, the supernatant is placed into ultracentrifuge tubs and spinned in the Beckman ultracentrifuge at 150,000 g (rotor SW41) for 45 minutes.

5) The sediment of SMP is very sticky, therefore carefully collect it and homogenize in a small glass homogenizer in a volume to give the final [SMP] of 10-20 mg/ml. Use the medium: 150 mM sucrose, KCl 75 mM, MOPS 5 mM, pH 7.2. A usual recovery is about 20-30% of the original protein concentration of mitochondria.

6) It is more reliable to use the Lowry method for determination of the protein concentration in SMP.

IMPORTANT. Because SMP undergo rapid oxidative damages when stored in the freezer at -80°C (see below), it is better to store them either in liquid nitrogen, or at -80°C in a tightly closed vessel filled with the Pirrogallol-treated nitrogen (to remove the remnants of oxygen). Do not store SMP at -80°C for a long time (more than a week).

Possible sources of errors and variations in estimated activities of the respiratory chain complexes

Estimations of the specific activities of respiratory complexes are commonly conducted for the diagnostic and research purposes on SMP prepared from mitochondria isolated from the cultured cells, usually lymphoblastoid cells derived from the patients' lymphocytes. Therefore I also describe the steps related to the cell culturing.

Steps and sources of errors.

Cell culture. Variations in the culture conditions may affect both the number of mitochondria per cell and the enzyme activities. For example: a) high glucose concentration, 20 mM and more, in the culture medium might diminish the total amount of mitochondria in cells; b) The presence of antibiotics (streptomycin) inhibit expression of mtDNA-dependent proteins, thus resulting in lower activity of the respiratory complexes. I have also noticed that the number of mitochondria in cultured cells is a phenotypic phenomenon. For each cell line, the number of mitochondria is different. Therefore it is impossible to utilize strict quantitative comparisons of two cell lines of different origin. For example, human lymphoblastoid cells from normal or diseased individuals may have large quantitative diversity in mitochondrial functions, but qualitatively are similar (Panov et al., 2007).

Isolation of mitochondria. If the harvested cells were disrupted with digitonin, the latter might damage mitochondria or prepared SMP, particularly, if the mitochondria were rich in cholesterol as is in the case of some cancer cells. Traces of digitonin may also interfere with the enzyme activity via changing interactions of SMP with ubiquinone or its derivatives. Contamination of mitochondrial fractions with the non-mitochondrial proteins also will affect the calculated activities.

Protein determination. Incorrectly determined protein strongly affects the calculated activities.

Preparation of SMP. Normally, the so called EDTA-SMP have the inside-out orientation of the vesicle's membrane as compared with mitochondria. If EDTA was not present in the sonication medium, than the proportion of the inside-out SMP could vary significantly, and thus affect the estimated activities. It is unacceptable to measure the activities using the frozen-thawed mitochondria. The results will be much more reproducible if after sonication the SMP were spinned at 16,000 g for 10 min. to remove large particles. Variations in the sonication conditions can also produce SMP of different size and "quality", or through temporary temperature rise and oxidation (air bubbles).

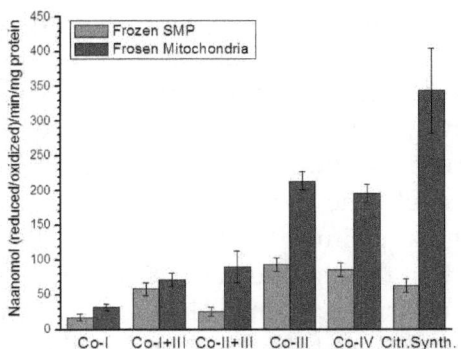

Figure 5.1. Effect of 14 days storage of SMP and Mitochondria at -80oC on enzyme activities of SMP from human lymphoblastoid cell mitochondria. Mitochondria were isolated from cultured human lymphoblastoid cells. Part of mitochondria was used to prepare SMP. Both mitochondria and SMP were stored at -80oC. After 2 weeks of storage, mitochondria were thawed and SMP prepared. The enzyme activities were measured in the prepared and thawed SMP. Light-gray columns represent stored SMP, dark columns represent SMP from stored mitochondria.

Storage of SMP. SMP have very large surface area in comparison with the whole mitochondria, therefore oxidative damages of the membrane and proteins are very likely during the storage. It is known that the solubility of Oxygen increases with the fall of the temperature. In addition, the mobility of H^+, OH^- and O_2 radicals is several orders higher in ice than in liquid water. From your own experience you know that a chicken stored in the freezer often acquires specific smell of rancid fats.

Figure 5.1 compares activities of the respiratory chain complexes and citrate synthase measured in SMP prepared from the freshly isolated mitochondria, and then stored at -80°C for 2 weeks, and in SMP prepared from the same mitochondria, but which were stored whole for 2 weeks at -80°C. The figure 5.1 shows that the activities of all respiratory complexes and Citrate Synthase, in particular, are very sensitive to oxidative damage. After the storage of SMP at -80°C for 2 weeks, the activities of most enzymes drop significantly to a different degree.

It was published in the literature that the activities of the respiratory complexes vary widely between the human individuals (Rustin et al. 1994). Because of that, it was impossible to use regular statistic methods. Therefore, some authors use the ratios of enzyme activity/Citrate synthase (CS) activity. However, Fig. 5.1 demonstrates that because of high sensitivity of CS to oxidative damage, normalization of the activities of respiratory chain complexes in SMP to the activity of CS is obsolete. When comparing SMP from HLB cells of different individuals prepared from mitochondria stored for the same period of time, I did not find large differences between the individuals, as was claimed by Rustin et al. (1994).

It is evident, that the storage at –80°C does not prevent oxidative damage of mitochondrial proteins. Therefore, do not store SMP at -80°C, or store them without thawing under nitrogen or argon. It is much better to store frozen mitochondria under nitrogen or in liquid nitrogen, and prepare SMP just before the measurements.

Conditions of the enzyme activity measurements. There are large variations in the conditions of biochemical measurements. This is particularly true in respect of the Complex I activity. Some people use freeze-thawed mitochondria instead of EDTA-SMP, various derivatives of CoQ, such as Q1, Decylubiquinone (DB), and (generally) too high concentrations of both DB and NADH.

Figure 5.2 shows that in fact there are rather complicated relations between the Complex I activity and concentration of DB, particularly in cancer mitochondria. These measurements reflect Complex I activity in respect of the hydrophobic part of the complex. In addition, it is possible to measure the affinity of the enzyme to NADH, which reflects the hydrophilic part of the complex, by determining the K_M of the Complex I for NADH at the optimal concentration of DB determined as shown in Fig. 5.2.

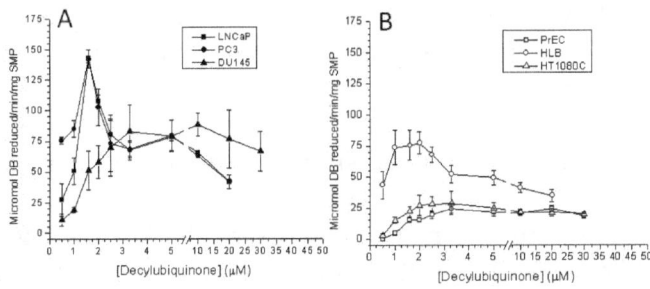

Figure 5.2. Dependence of Complex I activity on concentration of Decylubiquinone (DB) in submitochondrial particles from normal human prostate epithelial cells (PrEC), normal human lymphoblastoid cells (HLB), human prostate cancer cell lines PC-3, DU145, LNCaP, and human fibrobsarcoma HT1080 cells. Incubation conditions: 250 mM sucrose, 2 mM EDTA, 10 mM Tris-HCl, pH 7.4, NADH 1 mM (freshly prepared), KCN 10 mM, rotenone 20 μg, various concentrations of DB, SMP 0.15 mg. Reduction of DB was measured at 272 minus 247 nm (extinction coefficient 8 mM^{-1}cm^{-1}). After incubation for 5 min at 30ºC, the reaction was started by addition of 1 mM NADH. **Panel A:** the rate of DB reduction by SMP from LNCaP, PC-3, and DU145 cells. **Panel B:** the rate of DB reduction by SMP from PrEC, HLB, and HFS 1080C cells (Panov & Orynbayeva, 2013).

One should also be very meticulous in respect of controls. Non-specific activities vary significantly from sample to sample, particularly for complex I + III, and II + III. The activity of Complex II is inhibited in SMP.

CHAPTER 6

Methods to Study Mitochondrial Functions

General considerations.

When you plan to isolate mitochondria from any organ or tissue, or cultured cells, the first question is - how to study these mitochondria? What functions and methods are best to characterize mitochondria from this particular source? Will the "conventional" methods be enough?

When you just begin to ask yourself these questions, you instantly realize that there is no simple answer to either of them. Let us begin with the methods. What methods are the best? What function or functions the method will reflect? Are the methods that are suitable for the liver mitochondria will be good to study heart mitochondria? The simplest suggestion would be to use methods what most people use, say, to study respiration you can use succinate + rotenone, "as everybody does". This is an example of the most "conventional" substrate mixture. A very stupid approach to the above questions, as you will very soon realize yourself.

By the way, did you notice that during the last 5-6 years there were published few papers describing actual measurements of mitochondrial respiration or membrane potential? And this is in spite of the fact that mitochondria are one of the "hottest" issues in biomedical research. I think that partly this was the result of complications with the water quality described in the Chapter 1. But also, because many researchers "feel" that "conventional" methods in most cases are not only uninformative, but erroneous. And no alternative methods to study the principal mitochondrial function – respiratory activity were offered. In spite of the fact, that new instruments to measure oxygen consumption, alternative to the classical Clark electrode, have appeared on the market, for example, oxygraph that uses fluorescent probes sensitive to oxygen concentration; Oroboros oxygraph, etc., no new methodology in terms of substrate specificity was offered.

Here, I want to stress the more general problems, which are of great importance for understanding mitochondrial and organ's energy metabolism. Without taking into consideration metabolic differences between organs you will not understand the methodological problems in mitochondrial studies. Currently, animals are widely used to model human diseases. Huge amount of money were spent uselessly because of "conventional" methodology, "conventional" methods, and simply narrow-mindedness and ignorance.

Let us begin with the "simple" questions: Are mitochondria from different organs and tissues similar, or different? Why the particular model of a disease, even with clear involvement of mitochondriopathy of genetic origin, can be reproduced on one animal species but not on the others? Basically, these are fundamental questions, which have no simple and quick answer. Unfortunately, in USA (and perhaps in other countries as well) fundamental questions are unpopular because nobody gives money to solve 'impractical" problems. The grants are given to resolve "really practical" problems that would "quickly" result in new methods of prevention and treatment of diseases. Seemingly the bureaucrats at NIH and other agencies distributing finances do not know the old saying that "there is nothing more practical, than a good fundamental theory".

In order to find answers to the above questions, I will begin with general considerations regarding the most fundamental functions of mitochondria –production of energy and related metabolism of substrates.

Relationships between mitochondrial functions and the functions of related organs or tissues

Every organ is a complex, specifically organized part of an organism with separate specific functions. These are the liver, heart, kidney, brain, spleen, etc. All organs consist of different cell types, some of them may be present in other or even all parts of the body and organs, and have specific functions of their own distinct from other cell types. These are connective tissue, epithelial cells, glial cells, fat cells, etc. Therefore, we can assume two extreme possibilities: 1. Mitochondria are autonomous subcellular organelles, which have specific functions that are the same in any organ or tissue; 2. Mitochondria from different organs and tissues are distinct and their structure and functions are determined by the properties of the host cells. As we accumulate more and more facts about mitochondria, the truth is somewhere between these two extremes, or both statements can be valid under certain conditions. It is like with the light, it may be a wave or a particle, depending how you study.

Let us start from the "beginning", by defining what Energy Metabolism is, and how it relates to other types of metabolisms (metabolism of carbohydrates, fats, etc.) through mitochondria. All cellular functions in higher animals require energy in the form of ATP, NADPH, pH and concentration gradients of metabolites, all of which are generated directly or indirectly by mitochondria. This "useful" energy can be generated only if much more energy is released in irreversible processes accompanied by dissipation of more energy and an increase in the system's Entropy. Mitochondria are devices, which produce "useful" energy by utilizing part of the Free Energy released during the burning of hydrogen derived from the different food sources: carbohydrates (glucose), fats (fatty acids) and proteins (aminoacids). This General Statement presumes that basically all mitochondria have to be the same in all organs and tissues. But let us consider the whole body's Energy Metabolism and mitochondrial functions from the functional and metabolic points of different organs.

A person, who had a good supper and slept soundly over the night, in the morning has full reserves of carbohydrates in the form of blood glucose and glycogen in the liver and muscles, some amount of aminoacids in the blood, and fats as a fat tissue and fatty acids bound the to blood's albumin and transport proteins. How long each of these storages will last as the source of hydrogen in a person who performs a mild work, say, a quiet walk? And what organs will be consuming these storages?

Let us first consider carbohydrates. The content of glucose in the blood is close to 1000 mg per 1 L that is 5 grams in an average person with 5 liters of blood. This amount of glucose will be consumed by erythrocytes alone just in 20 – 30 minutes, and in a much less time, if glucose is also consumed by the brain, spinal cord, heart, skeletal muscles, kidneys and liver. Usually, the largest storage of glucose in the liver as glycogen is about 100-120 gram (Campbell et al. 2006). Only glucose from the liver's glycogen can be shared with other organs. We should also keep in mind that fat tissue also consumes large amount of glucose. This is because in the fat cells there is a constant cycling of fatty acids: first triglycerides split into free fatty acids and glycerol, and then the same fatty acids become reconnected to a new glycerol, to form a triglyceride, while the "old" glycerol is catabolized.

As a result, after about 2-3 hours of mild work, all reserves of carbohydrates (glucose) will be depleted. Therefore, in order to maintain the physiologically constant level of glucose in the blood, the liver metabolism must switch from glycolysis to gluconeogenesis. The constant level of glucose in the blood is maintained for the sake of normal functioning of cells, which have absolute requirement for glucose: erythrocytes, nerve tissue and fat tissue.

The storage of amino acids is very small, and during gluconeogenesis and neuronal activity the pool of amino acids must be constantly replenished by transamination of α-ketoglutarate, pyruvate, or proteolysis of proteins. Fat is the largest and most energetically rich storage of hydrogen.

Thus, the constantly hard working organs, such as heart and kidneys, MUST utilize fatty acids as the major source of hydrogen almost at all times. There is a saying - "fats are burned in the flame of carbohydrates". As you will see from our further discussions, the correct statement should be "fats are burned in the flame of carbohydrates and proteins (amino acids)". This is demonstrated by experiments described below.

Table 6.1. Relative Caloric Values and Storage Amounts of major Sources of Mitochondrial Substrates

Source of Energy (Caloric value – kcal/g)	Storage amount (time of consumption)
Carbohydrates (CV = 3.81): Blood glucose & Glycogen	4-5 grams (20-30 min) 100-120 grams (2-4 hrs)
Amino acids (CV = 3.12)	Released during catabolism of food or tissue proteins.
Acyl Fatty Acids (CV = 9.3)	Fat (Kilograms)

Metabolic states of mitochondria.

Before we start considering the respiratory activity of mitochondria with various substrates, I have to remind the reader the basic metabolic states, which have different rates of oxygen consumption and different functional meanings.

The "State 4", or resting respiration, corresponds to mitochondria oxidizing substrates when the membrane is fully energized, but the mitochondria do not perform useful work. The initial State 4 is usually designated as State 4_0 to distinguish from other similar metabolic states. In State 4_0 adenine nucleotides (ADP and ATP) outside of mitochondria are absent, and membrane potential is at maximum. The State 4_0 respiration rate is determined by the intrinsic proton conductivity of the inner membrane, which dissipates the $\Delta\Psi$ (Brown et al. 1994). I have evidence that the rates of the State 4 respiration depend also on the reverse electron transport, which is the energy-dependent process (Panov et al, 2012).

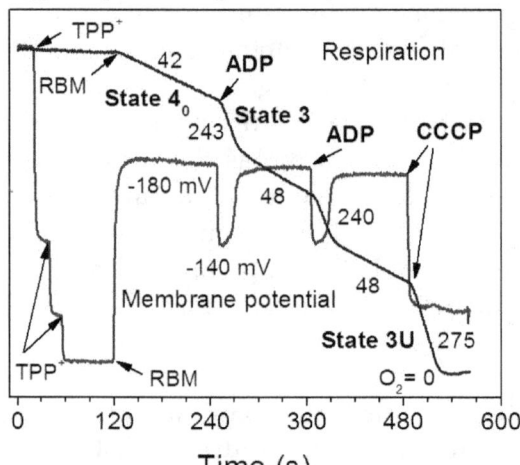

Figure 6.1. Oxygen consumption and membrane potential in rat brain (Sprague Dawley) mitochondria oxidizing pyruvate + glutamate + malate. Incubation conditions: The incubation "Full medium" (see Table 6.2) contained 10 mM glutamate + 2.5 mM pyruvate + 2 mM malate, final volume 0.65 ml. Additions: RBM 0.3 mg, ADP 150 μM, CCCP 0.4 μM. The numbers show the rates of O_2 consumption in nanomol O_2 per minute per 1 mg mitochondrial protein. The numbers at the the membrane potential trace show the values of calculated $\Delta\Psi$ in mV as determined by the TPP+- sensitive electrode. The details of the methods will be described later in the corresponding sections.

Addition to mitochondria of ADP stimulates oxidative phosphorylation, which is conventionally designated as the "State 3" respiration. As a result, membrane potential drops by 20-40 mV, and respiration increases. The degree of the $\Delta\Psi$ decline depends on the ability of mitochondria to stimulate respiration. The State 3/State 4 ratio of the respiratory rates is called the Respiratory Control Ratio

(RCR), the details of its calculations and meanings will be discussed later. In figure 6.1 the RCR ratio is 5.8.

Many researchers believe that high RCR evidences the high quality of mitochondria, and a decrease in RCR is interpreted as a sign of uncoupling and diminution in the quality of mitochondria. However, RCR alone is meaningless, and has only significance when the respiratory rates are known.

With some substrate mixtures, a decline in RCR does not mean deterioration of mitochondria, but simply reflects an increase in the State 4 associated with increased reverse electron transport. Decreased RCR may be also associated with the inhibited phosphorylation of ADP. After mitochondria finish phosphorylation of ADP, the $\Delta\Psi$ returns back to normal, and respiration declines back to state 4, which in this case is designated as State 4_1. In State 4_1 in the incubation medium ATP and ADP are present, which interfere with mitochondria through the adenine nucleotide translocase (ANT). In ideal (see Fig. 6.1), the rate of State 4_1 is similar or close to State 4_0. The presence of the non-specific ATPase activity may increase State 4_1 and thus influence the RCR values calculated as State 3/State 4_1. Thus only by analysis of the respiratory rates in each metabolic state, the changes in RCR may be interpreted correctly.

Addition to mitochondria of a protonophore, in our example that was CCCP (see Fig. 6.1), results in the inhibition of oxidative phosphorylation due to dissipation of the membrane potential. Therefore, respiration increases, and thus phosphorylation and respiration become "uncoupled". This metabolic state of mitochondria is designated as the "State 3U (uncoupled)". However, stimulation of respiration by uncouplers often does not occur. On the contrary, far too often respiration becomes strongly inhibited and is substrate-dependent. If mitochondria oxidize succinate in the absence of rotenone, respiration often becomes inhibited by oxaloacetate (OAA). Vinogradov with colleagues have shown that in the de-energized mitochondria the affinity of SDH to OAA may increase 10 fold (Vinogradov et al. 1972). With glutamate + malate, the inhibition of respiration in the presence of uncoupler occurs because transport of glutamate is energy-dependent. This inhibition will depend on the type of mitochondria, a substrate mixture, and the animal's phenotype.

Mitochondrial respiratory rates depend on the organ of origin, substrate type and substrate combinations

The figures below show polarographic traces of oxygen consumption and membrane potential changes (measured with a TPP^+-sensitive electrode) by rat heart mitochondria (RHM). The mitochondria were isolated from the heart of a young male Sprague Dawley rat by the method described in Chapter 4. The conventional approach to investigate the respiratory activity of RHM would require utilization of the respiratory substrates separately.

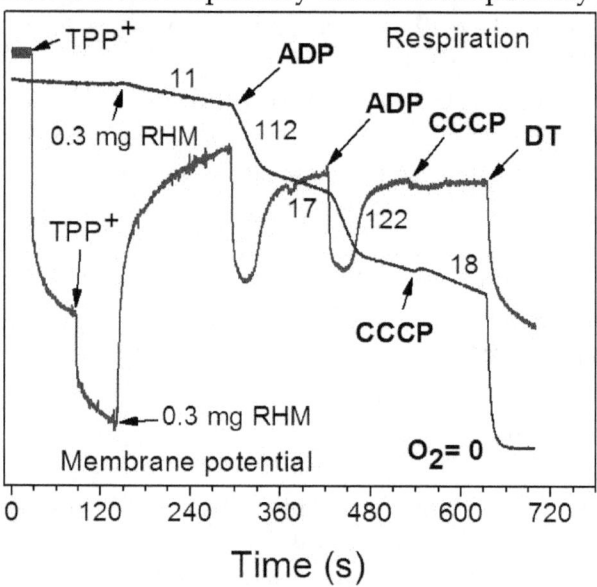

Figure 6.2. Rat heart mitochondria (RHM) oxidizing 10 mM glutamate + 2 mM malate. Incubation coditions were as in Fig. 6.1.

In Fig. 6.2 Dithionite (DT) was added in order to save time. Dithionite instantly consumes oxygen dissolved in the incubation medium and thus saves several minutes. The polarographic experiments are relatively long, therefore it is essential to save time, if the respiration after addition of an uncoupler is too slow. You cannot just stop experiment without knowing the position of the zero oxygen point. Therefore dithionite helps to save up to 30 – 40 minutes. This is essential if you work with mitochondria from diseased animals. After the oxygen zero point was achieved, quickly wash the electrode and the chamber with ample amount of water.

RHM have low State 4 and high rate of oxidative phosphorylation (State 3), but the state 3U is strongly inhibited.

Surprisingly, simultaneous measurement of $\Delta\Psi$ shows that membrane potential was not collapsed immediately upon addition of CCCP. As I have found in additional experiments (not shown) this was associated with the relatively low concentration of added CCCP. After second addition of the uncoupler the membrane potential begun to decline.

Although in the *in vitro* experiment RHM oxidizing glutamate + malate can maintain high rates of the state 3 respiration, this is not a physiological substrate mixture because normally amino acids are not the major substrates for the energetics of the heart.

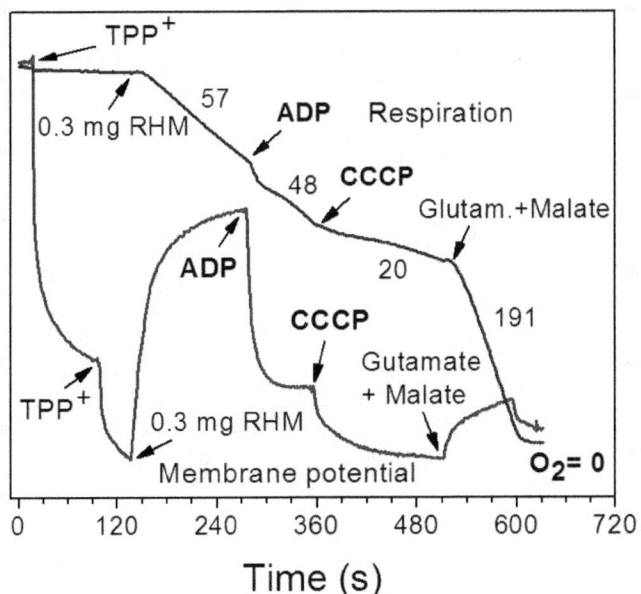

Figure 6.3. Respiratory rates and membrane potential for RHM oxidizing succinate in the absence of rotenone. Incubation conditions as in Fig, 6.1.

Many researchers use succinate in combination with rotenone. But in the cell there is no rotenone. Rotenone abolishes several physiologically and metabolically important events, which normally occur in a cell and mitochondria. The inhibition of succinate oxidation by oxaloacetate is not an artifact, but important mechanism of minimizing production of ROS by mitochondria (Panov A., et al. 2007). Fig. 6.3 demonstrates that succinate alone is a "bad" substrate because both oxidative phosphorylation and uncoupled respiration become strongly inhibited. But do not use rotenone! Instead, use the appropriate mixture of substrates.

You can see that as soon as ADP was added to RHM, the rate of oxidative phosphorylation rapidly became inhibited. The inhibition was further enhanced after addition of uncoupler (CCCP). The inhibition was released upon addition of glutamate + malate. First, you see that succinate alone as a respiratory substrate is a "bad" substrate. The inhibition was caused by oxaloacetate, which accumulated in mitochondria in the absence of rotenone.

Under physiological conditions, RHM oxidize predominantly fatty acids. However, Fig. 6.4 shows that although with palmitoyl-carnitine + malate RHM showed high rates of State 4_0 and State 3 respiration, uncoupling of mitochondria also cased inhibition of oxygen consumption. Thus, with the isolated RHM both succinate (without rotenone) and palmitoyl-carnitine do not provide high rates of respiration in all metabolic states.

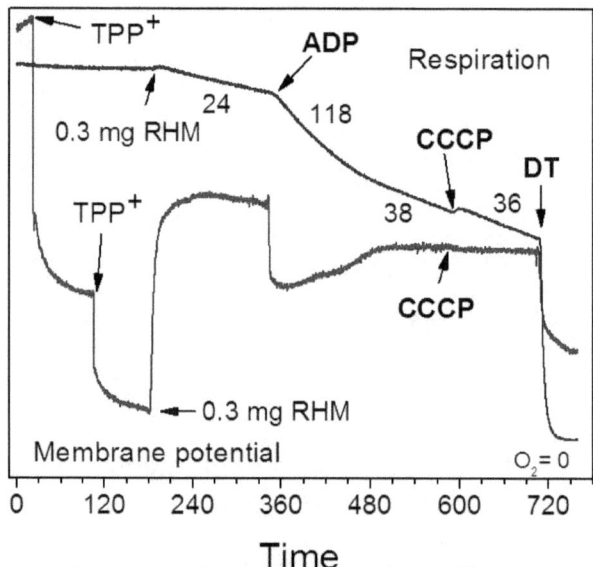

Figure 6.4. Respiratory activity and membrane potential in RHM oxidizing palmitoyl-carnitine 0.25 mM + malate 2 mM. Incubation conditions as in Fig, 6.1.

Figure 6.5 shows that when RHM were exposed to the mixture of two "bad" substrates (succinate and palmitoyl-carnitine), the respiratory activities were high in all metabolic states. Moreover, there was no inhibition of respiration in the presence of uncoupler. The State 4 respiration was unchanged as compared with the succinate alone (Fig. 6.3) and increased 2.5 fold when compared with palmitoyl-carnitine alone (Fig. 6.4.). The State 3 oxidation of the substrate mixture was 40-60% higher than with glutamate + malate or palmitoyl-carnitine.

High rates of the State 4 and the State 3 respiration were also observed when RHM oxidized palmitoyl-carnitine + glutamate + malate (Figure 6.6). Although CCCP increased oxygen consumption, soon it became inhibited to the state 4 level. The inhibition was probably associated with the fact that transport of both glutamate and palmitoyl-carnitine into mitochondria is energy dependent. Importantly, the increased State 4 respiratory rates with physiologically relevant substrate mixtures reflect the dramatically increased rates of ROS generation associated with the reverse electron transport, which I measured in separate experiments.

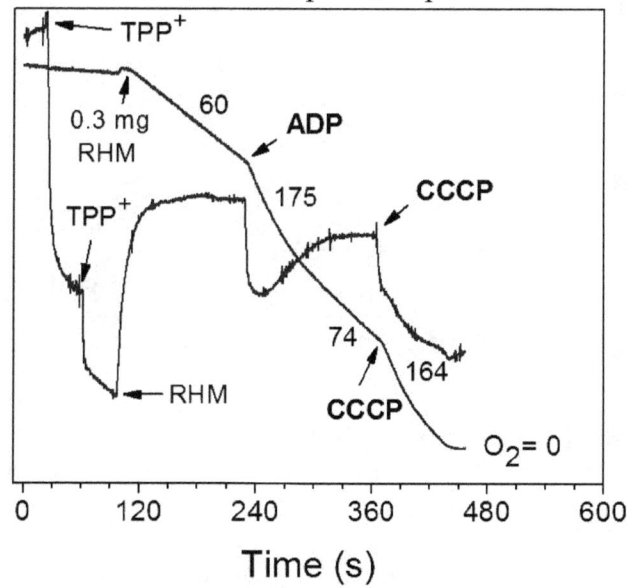

Figure 6.5. Respiratory activities of the RHM oxidizing succinate 5 mM + palmitoyl-carnitine 0.25 mM. Incubation conditions as in Fig, 6.1.

There is a close correlation between the rates of the State 4 respiration and ROS production by mitochondria (Panov et al. 2009, 2011). This suggests that both respiration and ROS generation were associated with the reverse electron transport. We will discuss this later in chapter 12 devoted to mitochondrial production of ROS. What is demonstrated in figures 6.1, 6.2 and 6.3, that the "conventional" use of substrates either makes no physiological or scientific sense (succinate + rotenone), or does not reflect physiological situation (glutamate + malate).

With Palmitoyl-carnitine as substrate, RHM may have high rate of oxidative phosphorylation (State 3 respiration), compared with glutamate + malate. However, upon addition of uncoupler, the oxidation of palmitoyl-carnitine becomes inhibited. It is important to remember, that glutamate alone is not a major physiological substrate for the heart.

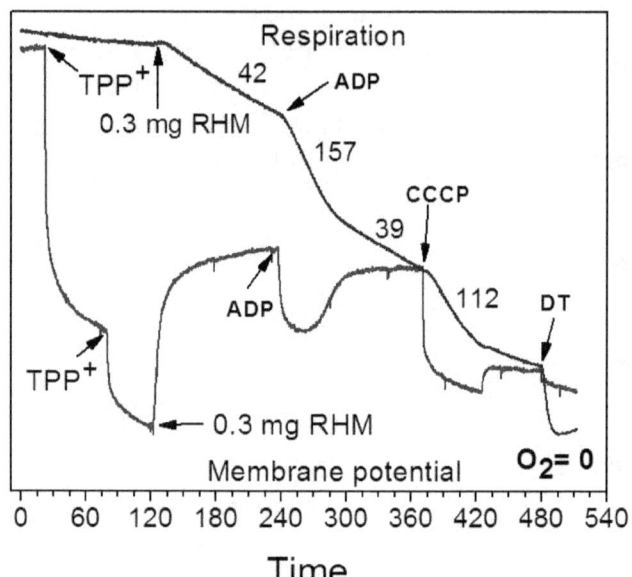

Figure 6.6. Respiratory activities of RHM oxidizing glutamate 10 mM+ malate 2 mM + palmitoyl-carnitine 0.25 mM. Incubation conditions as in Fig, 6.1.

From the Common Sense, it seems clear that mitochondria in the cell never oxidize a single substrate. The reducing equivalents, i.e. protons and electrons, enter respiratory chain from different sites and stages of metabolic pathways. The scientifically sound method to study mitochondrial functions suggests that the researcher must utilize the physiologically relevant mixtures of substrates, which are different for different tissues. However, this issue was not yet thoroughly studied. Although physiologists for a long time know that heart, for example, utilizes a mixture of fatty acids and glucose for energy production, mitochondriologists continue utilizing succinate + rotenone or single substrates (glutamate + malate). My publications were the first to stress the significance of substrate mixtures to study the brain and spinal cord mitochondria (Panov et al. 2009, 20011).

The reasons why the methods to study liver mitochondria still dominate. Historically, it happened that most of our current knowledge about mitochondrial functions was obtained in experiments with the liver mitochondria. Other popular objects for early mitochondrial studies were beef heart mitochondria (BHM), particularly submitochondrial particles, because from one beef heart you could isolate a "ton" of BHM and store them frozen or as SMP for a long time. I remember how in 1971, when I was working at Lars Ernster's laboratory at the Kungliga University in Stockholm as a WHO scholar, the whole laboratory staff was mobilized to participate in the large scale isolation of mitochondria from the huge beef hearts, and then preparation of SMP. In two days the freezer was filled with dozens of vials with SMP, which we than used for months to study transhydrogenase and other enzymes. Of course, frozen mitochondria were unsuitable for any kind of studies of either mitochondrial functions or mitochondrial enzyme activities. Unfortunately, there are papers where the authors did just that.

For several decades researchers were busy with the fundamental problems of mitochondriology, such as the mechanisms of electron transport, ATP synthesis, ion transport, etc. In this respect liver mitochondria were a perfect object for research because of the simplicity and low price of isolation. Until the major fundamental problems of electron transport and oxidative phosphorylation were resolved, much less attention was directed towards the physiological aspects of mitochondrial functions, or involvement of mitochondria in diseases.

Liver as an organ, and thus liver mitochondria, stand apart from other organs and tissues, such as heart, brain, skeletal muscle, kidney, etc. But this important fact was not appreciated by many researchers, and some of them still regard liver mitochondria as a standard for evaluating of mitochondria from other organs. Unlike liver mitochondria, mitochondria from other organs and tissues do not depend on the metabolic state of the body to such a degree as the liver mitochondria. Because it is one of the major functions of the liver to support metabolic homeostasis for other organs and tissues, the liver mitochondria must be capable to switch from one type of substrate to another in accord with the changing metabolic demands. The overall metabolic state of the liver is determined by two parameters: the type of food, and the time passed from the last feeding. Our discussions of metabolism and liver functions consider omnivorous organisms, such as humans, pigs, rats and mice. The livers of these organisms are capable to convert carbohydrates into fatty acids, fatty acids into carbohydrates, and amino acids into carbohydrates, depending on the type of food. But economical grounds and apparent "simplicity" of isolation and study were the main reasons why most experiments were conducted on mitochondria from rat and mouse liver. I and my colleagues Valentin Vavilin and Vladimir Solov'ev studied the dynamics of transition of the liver's metabolism from the fed to the starved state. We have found large differences between the rat species in this respect (Panov et al. 1991). One strain of rats , the Wag rats, showed no difference between the two metabolic states of the liver (levels of ATP, acyl-CoAs, Phosphate Potential dynamics, etc), and the properties of LM between the fed state (4 hours after feeding) and the starved state (12 hours after feeding) were also similar.

Significance of the Metabolic (Genetic) Phenotypes. Let us consider how in a human body substrate utilization by mitochondria in different organs might depend on the metabolic state of an organism during transition from the "fed" to the "starved" state.

For millennia, the humans' metabolism was determined by the environmental conditions. In the far North the people's diet consisted predominantly of fat and protein, in the Tropical areas, the people's diet consisted predominantly of carbohydrates, often limited in essential amino acids, which promoted cannibalism in some regions. As a result, people of different racial or ethnical origin acquired substantial differences in metabolism of fats, carbohydrates, etc, which I designate as metabolic phenotypes. We mention all this because in experimental mitochondriology we always have to keep in mind that animals, which are the main objects for modeling human's diseases, also have significant differences in organization and regulation of metabolic processes and thus mitochondrial functions.

CHAPTER 7

Incubation Buffers and Mitochondrial Substrates

Incubation media

In the literature you can find a large number of incubation media with different compositions. It is impossible to discuss all of them critically. There are basically three types of media: sucrose based, potassium based, and sodium based media. The Na-based media are physiological solution (0.9% NaCl), Ringer solution and Krebs-Ringer solution. They are generally used for incubation of cells and manipulations with them. These media are completely unacceptable for incubation of mitochondria.

Table 7.1. Composition of some Na-based buffers.

Components	Ringer's solution (mM)	Krebs-Ringer solution (mM)
NaCl	111.00	25.00
KCl	5.63	2.50
CaCl$_2$	3.25	2.00
NaHCO$_3$	100.00	25.00
NaH$_2$PO$_4$	-	1.25
MgCl$_2$	-	1.00
Glucose	-	25.00
Tonicity	340.0 mOsm Slightly hypertonic	115.5 mOsm Strongly hypotonic

Ringer solution is used sometimes in Medicine, but originally was designed for experimental perfusion of the frog hearts. The Krebs-Ringer solution is often used for incubation of cells. These media are completely unacceptable for incubation of mitochondria.

The sucrose based medium, usually 250 mM sucrose, with buffer, inorganic phosphate, and MgCl$_2$, lately is rarely used to study mitochondrial respiration. This medium does not provide conditions for maximal rate of oxidative phosphorylation. The latter requires the presence of at least 20 mM KCl. The table 7.2 shows the composition of the sucrose-based medium, which contains 20 mM KCl plus 5 mM

K+, if inorganic phosphate is KH_2PO_4, or about 10 mM K+, in case of K_2HPO_4. The total [K+] may be about 30-35 mM, if you also consider the amount potassium added as KOH when adjusting pH with 20 mM MOPS.

Table 7.2. The Sucrose-based medium for incubation of mitochondria.

Components	Mol. Weight	1 L
Sucrose 200 mM	342.3	68g 460 mg
KCl 20 mM	74.56	1 g 490 mg
MOPS 20 mM, pH 7.2	209.26	4.2 g
K_2HPO_4 5 mM	174.18	870.9 mg
OR KH_2PO_4 5 mM	136.2	681 mg
$MgCl_2$ *$6H_2O$ 5 mM	203.3	1g 016 mg
EGTA 1 mM	380.4	380.4 mg
Tonicity: ≈300 mOsm.		

This medium provides high rates of oxidative phosphorylation with liver mitochondria oxidizing glutamate + malate or succinate. However, the sucrose–based medium does not provide optimal conditions for the State 3 respiration with brain, heart or kidney mitochondria. It is a characteristic of liver mitochondria that they have much lower, than mitochondria from other organs, sensitivity to the composition of the incubation media, or whether liver mitochondria were isolated with or without BSA (see later discussion).

Table 7.3 shows the composition of the incubation medium, which is the best to study mitochondrial respiratory activities and other energy-dependent functions, such as $\Delta\Psi$ measurements, generation of ROS, etc. I used this medium for the heart, brain, spinal cord, skeletal muscle, kidney mitochondria and mitochondria isolated from the cultured cells. Because I always isolated liver mitochondria for the control purposes, to adjust equipment, I also incubated liver mitochondria in this medium with perfect results.

Let us discuss the composition of the incubation medium presented in Table 7.3 and what each component does to make this medium perfect for most, but not all, mitochondrial functions and types of mitochondria.

1). The incubation medium, which I designated as "Full" medium, has 120 mM KCl as the major ionic and osmotic component, which is close to the K^+ content in the cell.

Table 7.3. "Full" incubation medium for mitochondria.

Components	Mol. Weight	500 ml
120 mM KCl	74.56	4g 474
10 mM NaCl	58.44	292 mg
2 mM $MgCl_2$ $6H_2O$	203.3	203 mg
2 mM KH_2PO_4	136.09	mg
OR K_2HPO_4	174.18	174.18 mg
20 mM MOPS, pH 7.2	209.3	2g 094
1 mM EGTA (Ca-free)	380.4	190.2
0.7 mM $CaCl_2$	111.1	mg
$CaCl_2$ $2H_2O$	147.0	51.45 mg
($\approx$1 μM [Ca-free])		
Tonicity: $\approx$300 mOsm.		

2). The medium also contains 10 mM NaCl. This corresponds to the cytosolic contents of Na in the excitable cells, such as neuronal and muscle cells, as well as kidney. Mitochondria from these organs have the Na^+/Ca^{2+} antiporter (liver mitochondria have H^+/Ca^{2+} antiporter), which help mitochondria and cells to tune up precisely the mitochondrial $[Ca^{2+}]_{Free}$ Ca^{2+}.

3). $MgCl_2$ is essential component of the cytosol, and most of magnesium is bound to ATP and proteins. The concentration of $[Mg^{2+}]_{Free}$ is about 0.5 mM and is highly buffered. Mg^{2+} is necessary for maintaining structure of some proteins (see Commentary about EDTA). The presence of Mg^{2+} is also essential for correct kinetics of oxidative phosphorylation. Under conditions of a cell, ATP is anion with 4 negative charges (ATP^{4-}), ADP - as ADP^{3-}.

Most of ATP^{4-} is in the ATP^{4-}-Mg^{2+} complex, whereas most of ADP^{3-} is free. Because the ATP/ADP translocase (adenine nucleotide translocase –ANT) binds only free ATP^{4-} and ADP^{3-}, in the presence of Mg^{2+} the concentration of free ATP^{4-} is negligible, and this helps mitochondria to generate high ATP/ADP ratios. If Mg^{2+} was omitted in the incubation medium, the unbound ATP would compete with ADP for binding on ANT and thus change the kinetics of oxidative phosphorylation. There are also a number of related consequences of the $ATP^{4-}-Mg^{2+}$ complex formation, which serve to enhance the specificity of the enzyme-substrate interactions by enhancing the binding energy. First, the Mg^{2+} ion neutralizes some of the negative charges present on the polyphosphate chain, reducing nonspecific ionic interactions between the enzyme and the polyphosphate group of the nucleotide. Second, the interactions between Mg^{2+} ion and oxygen atoms in the phosphoryl group hold the nucleotide in well-defined conformations that can be specifically bound by the enzyme's active center. Mg^{2+} ions are typically coordinated to six groups in an octahedral arrangement. Third, the Mg^{2+} ion provides additional points of interaction between the $ATP-Mg^{2+}$ complex and the enzyme, thus increasing the binding energy. Such interactions have been observed in adenylate kinases bound to ATP analogs. (Biochemistry. 5th edition. Berg JM, Tymoczko JL, Stryer L. New York: W H Freeman; 2002).

4). Inorganic phosphate (Pi) is usually purchased either as a mono-potassium salt of phosphoric acid - KH_2PO_4, or as a di-potassium salt - K_2HPO_4. Before the pH meters were available, the mixture of these salts was used to achieve the necessary pH. It is still useful to know how to prepare the phosphate buffer. Using a pH-meter for adjusting the necessary pH, it does not matter which potassium phosphate salt you use. However, you should know the difference between mono- and di-phosphate salts when calculating the osmolarity of your buffers.

The exact total concentration of Pi in the cell is not known because much of it is bonded to ATP, proteins and chelated with calcium. The reasonable concentrations lie within 1-5 mM. I preferred to use 2 mM because I found that this is a optimal concentration of Pi, which provides maximal rates of oxidative phosphorylation with the heart or brain mitochondria. At higher than 5 mM, the experiments with calcium sequestration become questionable because too much Pi may be accumulated in the matrix and thus promote opening of the permeability transition pore.

5). The physiological pH in the cytoplasm is 7.2. Because MOPS has the pK_a of 7.2, the maximum buffer capacity is at the physiological pH. Tris-HCl or Tris-Base buffers have low buffer capacity at pH 7.2 because their pK values are at much higher pH. It is known that Tris forms complexes with Ag^+, and this will ruin the combination pH electrodes because Tris-Ag particles will clog the porcelain filter. For most experimental conditions 10-20 mM MOPS is more than enough.

6). The incubation buffer contains the calcium (as $CaCl_2$) 0.7 mM/EGTA 1 mM buffer, which provides the concentration of free calcium ($[Ca^{2+}]_{Free}$) close to 1 μM. I determined this with Fura 2 fluorescence (Panov, Scaduto, 1996). This concentration of $[Ca^{2+}]_{Free}$ is close to what exists in the mitochondria. The presence of calcium is necessary for stimulation of a number of mitochondrial dehydrogenases (for detailed discussion and related references see review by Brown G. C., 1992; and Panov & Scaduto, 1996).

EGTA (ethylene glycol tetraacetic acid) is a polyamino carboxylic acid, a chelating agent that is related to the better known EDTA, but with a much higher affinity for calcium than for magnesium ions and other divalent cations (Zn^{2+}, Cu^{2+}, Fe^{2+}). It is useful for making buffer solutions that resemble the environment inside living cells (Bett & Rasmussen, 2002) 1. Computer Models of Ion Channels. In Cabo, Candido; Rosenbaum, David S. Quantitative Cardiac Electrophysiology. Marcel Dekker. p. 48. ISBN 0-8247-0774-5), where calcium ions are usually at least several hundred times less concentrated than magnesium. As I have already mentioned, the cytosolic concentration of $[Ca^{2+}]_{Free}$ is about 0.1 μM, whereas $[Mg^{2+}]_{Free}$ is about 500 μM. That is the 500-fold difference.

Therefore EGTA is a perfect chelator to keep concentration of calcium in the isolation and incubation buffers at low concentration in the presence of high $[Mg^{2+}]$, and to prepare the Ca^{2+}/EGTA buffers giving controlled final concentrations of $[Ca^{2+}]_{Free}$. This is important because most chemicals, including sucrose, KCl, mannitol, and others, may be heavily contaminated with calcium. It is also important, that large amount of calcium can be released from the stores during preparation of tissue homogenates.

Because water and chemicals may be contaminated, besides Ca^{2+}, with other divalent cations, many researchers use, instead of EGTA, EDTA without consideration of important differences between the chelating properties of EGTA and EDTA. As a result, these chelators are often used indiscriminately. However, EDTA may lead to unwanted consequences, which may completely ruin experiments. Therefore, it is important to know the properties of EGTA and EDTA, use them with knowledge and care, always thinking about the consequences. For this reason I give more information about effects of EDTA on mitochondrial functions.

The unwanted effects of EDTA on mitochondrial functions. EDTA, ethylenediaminetetraacetic acid, also known as Versene, is a polyamino carboxylic acid and a colorless, water-soluble solid. Its conjugate base is named ethylenediaminetetraacetate. It is widely used to dissolve limescale. After being bound by EDTA, the metal ions remain in solution but exhibit diminished reactivity. EDTA is produced as several salts, notably disodium EDTA and calcium disodium EDTA. When EDTA is present in the isolation or incubation media, it will remove not only free Ca^{2+} and Mg^{2+} ions, but also other divalent cations from the mitochondrial high affinity sites. As a result, some proteins may change their conformation, and the inner mitochondrial membrane becomes permeable to H^+ (protons), and the membrane potential drops.

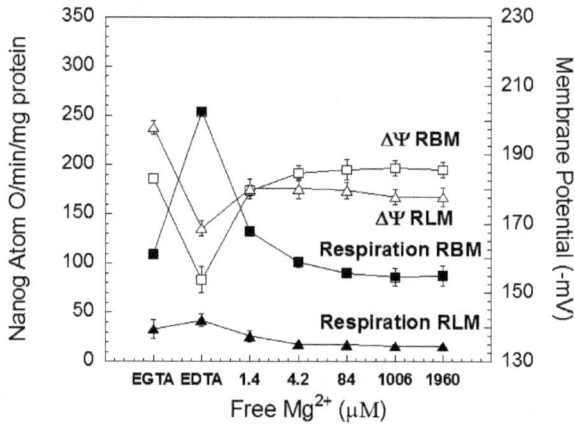

Figure 7.1. Effects of 1 mM EDTA and Mg^{2+} on the rates of the State 4 succinate oxidation and membrane potential in rat liver (RLM) and brain mitochondria (RBM). The mitochondria were incubated in the "Full" incubation medium with exception that $CaCl_2$ was omitted.

Figure 7.1 shows that when 1 mM EDTA was added to incubation medium, membrane potential dropped by 30 mV in RBM, and by 35 mM in RLM. Correspondingly, the State 4 respiration rates increased by 250% in RBM and by 40% in RLM. After 5 minutes, the aliquots of $MgCl_2$ were added and the concentrations of $[Mg^{2+}]_{Free}$ were calculated using the site: http://maxchelator.stanford.edu/CaEGTA-TS.htm. Both respiration and membrane potential returned to normal at concentration of $[Mg^{2+}]_{Free}$ 5 μM.

The decrease of $\Delta\Psi$ by 30-35 mV resulted in the complete inhibition of the reverse electron transport during oxidation of succinate or fatty acids and thus inhibited the associated production of reactive oxygen species. Similar changes in membrane potential and respiration I observed also in RBM oxidizing glutamate + malate and pyruvate + malate. This means that the effects of EDTA and Mg^{2+} were not associated specifically with a substrate, but were caused by changes in the proton conductivity of the inner membrane. Evidently, when EDTA removes Mg^{2+} from the high affinity sites, some protein (or proteins), we do not know yet which one, change conformation and open a channel for protons, and possibly for K^+. This problem still remains to be studied in full.

There is a good example how inconsiderate usage of EDTA led to erroneous conclusions. Sorgato et al. (1974) were one of the first to study production of ROS by brain mitochondria. They concluded that unlike heart and liver mitochondria, brain mitochondria do not produce H_2O_2 or superoxide radical (Sorgato et al., 1974). This erroneous conclusion delayed studies of ROS production by brain mitochondria for many years. Who would study this problem, if researchers from the well-known and highly respected laboratory headed by Dr. Azzi, have shown that brain mitochondria do not generate ROS? With time, however, people forgot about this work. When I begun to study brain mitochondria in 1997, there were several publications, which showed that RBM do produce both superoxide radical and hydrogen peroxide. Nevertheless, I was intrigued why Sorgato et al. (1974), a well respected team, could not find any sign of ROS production? So, I read the paper, and since I already had the data shown on Fig. 7.1, I instantly realized that because both the isolation and the incubation media contained 2 mM EDTA and no magnesium was added, the mitochondria were depleted of Mg^{2+} from the high affinity sites on mitochondria. The resulting drop in the membrane potential prevented the succinate-dependent reverse electron transport and thus ROS production. Sorgato et al. (1974), and other researchers in 1974, evidently did not know that EDTA has adverse effects on mitochondrial energization. The first publication on the ability of EDTA to increase the State 4 respiration was published by Cadenas & Brand in 2000. These authors studied the effects of EDTA and Mg^{2+} on the State 4 succinate oxidation by skeletal muscle mitochondria, and concluded that Mg^{2+}, but not purine nucleotides (GDP, ATP, ADP), control the proton conductivity of the inner mitochondrial membrane. Cadenas and Brand (2000) did not study changes in the membrane potential, or used other than succinate respiratory substrates. Here I have to mention, however, that ADP also controls the conductivity of the inner mitochondrial level to H^+ and K^+ in the presence of Mg^{2+} at the level of ANT (Panov et al. 1980).

IMPORTANT. The above data on the effects of EDTA, EGTA on mitochondrial functions call for very careful and thoughtful use of these chelators. If, for any reason you have to use EDTA at 0.5 or 1 mM concentration, for example to eliminate contamination of water or chemicals with Cu^{2+}, Zn^{2+} or Fe^{2+} ions, add to the incubation medium 2 mM $MgCl_2$. This will essentially eliminate the adverse effects of EDTA.

Substrates for mitochondrial respiration

During the studies of mitochondrial functions, you have to add substrates for mitochondrial respiration. I usually add substrates before addition of mitochondria. Sometimes, you have add them in the course of an experiment. Therefore, you have to prepare the stock solutions of substrates, which you can store in a freezer. Some substrates, however, are unstable and have to be prepared just before the experiment. The concentrations of substrates in the incubation medium depend on the K_M of the substrate transporter for this chemical. Some substrate transporters, such as tricarboxylate transporter, have a low specificity and thus can transport several different substrates. Some substrates, can be transported by a number of different transporter proteins. The general rule is that in order to avoid limitation of the maximum rate of respiration by the substrate's transport mechanism, the concentration of the substrate have to be several times higher than K_M for the transporter. For a number of substrates, such as pyruvate, glutamate and some others, malate is added to facilitate the exchange transport mechanism, or facilitate the malate-aspartate shuttle. Malate alone is oxidized very slowly. The suggested concentrations of substrates for mitochondria are presented in Apllication 2 at the end of the book.

Commentary 1. To prepare the stock solutions of substrates, dissolve the chemicals using either the incubation medium, or 10 mM MOPS with pH adjusted to 7.2 with KOH. The stock solutions of substrates should be of high enough in order to add them to the incubation chamber in a minimal volume. I preferred to add substrates in 10 μl volume per 1 ml. Therefore, the stock solutions must be 100 fold more concentrated. For the substrates that you can store in dvance, it is better to make the stock solutions in a relatively large volume (10-20 ml). This will minimize errors during preparation of the solutions and easier to adjust pH. Substrates as the Na or K salts usually do not require adjustment of pH. Substrates as acids require adjustment of pH to 7.2. Divide the solutions into 1.5 centrifuge tubes as 0.5 -1 ml aliquots and store at -20°C or -80°C. Most substrates can be stored frozen for months. However, if you thawed a vial, use it just for few days. The exceptions are ketoacids: α-ketoglutarate, pyruvate, and oxaloacetate. These substrates must be prepared daily. The solution of L-palmitoyl-carnitine dissolved in 50% ethanol can be used for several days.

Commentary 2. When you add to the incubation medium an inhibitor, uncoupler or other chemical dissolved in ethanol, keep in mind that by adding 5 µl of 70% ethanol to 1 ml of medium, the final concentration of ethanol (MW 46.07; Density 0.7893) will be 60 mM! Therefore, try to use ethanol at lowest concentration possible, that is minimize concentration as a solute (I use 40 or 50% EtOH to dissolve L-palmitoyl-carnitine and some other chemicals), and the volume of addition. Similar considerations relate to other solutes, DMSO (dimethyl sulfoxide), for example.

Commentary 3. It is very difficult to make a $CaCl_2$ solution with precise concentration. This is because the actual amount of water bound with $CaCl_2$ is unknown. Therefore exact $[Ca^{2+}]$ should be veryfied with either atomic absorption spectroscopy or by H/Ca estimation using the pH method. In the Table below are given calculations for the special brand of $CaCl_2$ with 99.99% purity from Sigma, which is stored, closed tightly in a desiccator.

$CaCl_2$ (99.99%) 35.9% Ca^{2+}, 64.1% Cl	MW 110.99	10 mM	11.1 mg / 10 ml

The precise concentration of $CaCl_2$ is not used often. For most other purposes, such as preparation of the Ca/EGTA buffer in the incubation medium, I used the same vial of $CaCl_2$ x $2H_2O$, which I kept tightly closed and rapped by Parafilm.

Inhibitors, Uncouplers and Ionophores

When you prepare the stock solutions of inhibitors and other biologically active chemicals, try to minimize the volume of additions as much as possible and reasonable. Most of the biologically active compounds are hydrophobic and therefore must be diluted in ethanol or DMSO. Addition of ethanol has much less adverse effects on mitochondria; therefore, if possible, use ethanol instead of DMSO (see

Commentary 2). Remember, that most inhibitors are very toxic to humans. In order to avoid cross-contamination of the vials with inhibitors, use separate Hamilton syringes for each inhibitor. When preparing KCN solution do not adjust pH because HCN is highly volatile, and keep the vial with KCN tightly closed. Do not breath while you fill the syringe with KCN, and try to work in a well ventilated room. Because KCN is highly unstable, your stock solution and final concentrations are in a large excess over actual inhibitory concentration. Do not use KCN with buffers, which contain glucose because glucose neutralizes the inhibitor.

Commentary 4. The stock solutions of uncouplers FCCP (Carbonyl cyanide 4-(trifluoromethoxy)phenyl-hydrazone) and CCCP (Carbonyl cyanide m-chlorophenyl hydrazone) must be stored in dark glass vials at -20°C. When necessary, the working solutions are prepared in a small volume (1 ml) because it will deteriorate faster than the stock solution. The stock solutions also gradually deteriorate and the degraded chemicals become less effective and may inhibit respiration. Therefore, after about 1 year, prepare the new stock solution. Although FCCP and CCCP are very similar in their effects, the FCCP becomes inhibitory during titration faster than CCCP. During the long term storage, FCCP also deteriorates faster. Therefore I preferred CCCP over FCCP.

Commentary 5. Addition of 5 µl of DMSO to 1 ml of buffer, gives final concentration of 74 mM. At this [DMSO] blocks formation of hydroxyl radicals and inhibits many of microsomal cytochrome P-450 isoforms. DMSO should be used with great caution, particularly if it contains some toxin. DMSO very quickly penetrates through the unprotected skin. Rubber gloves are not 100% protective. Therefore, just be very careful.

Commentary 6. To minimize the final concentration of ethanol during addition of chemicals, I used the following trick. I usually prepared the storage stock solution of a chemical at 10 or 100 fold higher concentrations than the working stock solutions. For example: First, I prepared the storage stock solution of CCCP at 0.6 mM concentration using 70% ethanol; then, I prepared the working stock solution of 0.06 mM using 50% ethanol. Addition of 5 µl of this solution to 0.65 ml chamber gave me 0.46 µM CCCP, which was enough to cause maximum uncoupling in most situations. However, with some substrate mixtures brain and heart mitochondria may increase respiratory rate significantly, and therefore will require more

CCCP for full uncoupling. Also, BSA and some other chemicals can bind uncoupler or inhibitor and thus change the effective concentration.

Commentary 7. Some inhibitors stick to the surface of the chamber or cuvette, particularly made of plastic, rotenone for example. In practice, it is very difficult or impossible to remove rotenone by washing the chamber with ethanol. The plastic custom made chamber you cannot wash with ethanol because it penetrates into micro-cracks and destroys the chamber. Therefore, use disposable plastic cuvettes for Spectrophotometric and fluorimetric measurements. To remove inhibitors from the respiratory chamber, first wash the chamber with ample amount of 50% ethanol and water, add incubation buffer and a large amount (10-20 mg/ml) of thick suspension of liver mitochondria, and incubate for 3-5 minutes. Repeat this procedure twice. This will remove all traces of rotenone or antimycin. Though, I preferred to have a separate chamber for experiments, which require addition of rotenone.

CHAPTER 8

Polarography

The most common method to follow oxygen concentration changes in a solution is to use the Clark electrode, invented by Leland C. Clark. Traditionally, this method is called polarographic method, and is based on the reaction of reduction of oxygen on the polished platinum electrode: $O_2 + 4\ e^- + 2\ H_2O \rightarrow 4\ OH^-$ (see Figure 8.1).

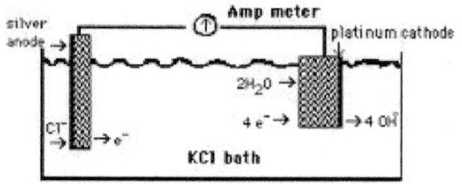

Measuring PO$_2$ (Clark electrode)

Figure 8.1. Principal electrical circuit of the Clark electrode.

The Ag/AgCl reference electrode serves as anode. The maximum performance of the Clark electrode is obtained at 0.65V with "+" applied to the platinum cathode. The construction of the combination Clark electrode is shown in Figure 8.2. The tip of the platinum should be polished and treated as described below. Combination electrodes are more convenient and may be of various sizes. To protect platinum from contamination with chemicals and protein, the electrode is protected by a thin Teflon or polyethylene membrane made from a common sandwich bag.

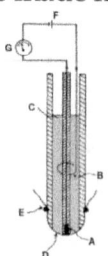

Figure 8.2. Principal construction of the combination Clark electrode. The electrode as a single unit combines platinum cathode and Ag/AgCl reference anode.

Solubility of Air and Oxygen. Solubility of air in water follows the Henry's Law, which states that the amount of air dissolved in a fluid is proportional with the pressure of the system. The solubility of oxygen in water is higher than the solubility of nitrogen. Air dissolved in water contains approximately 35.6% oxygen compared to 21% in air.

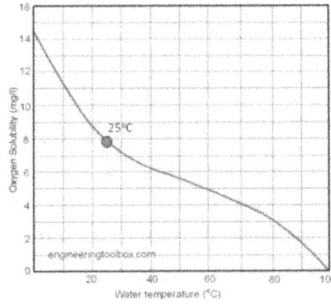

Figure 8.3. The solubility of oxygen strongly depends on the temperature. From Wikipedia

For the purpose of convenience and cost reduction, the measurements of mitochondrial respiratory activity are usually performed at room temperature (25°C). From the figure 8.3 one can see that at this temperature the solubility of O_2 is very sensitive to changes in the temperature. Therefore, for getting more reproducible results, the reaction chamber must be maintained at constant temperature. For this it is better to use a circulating thermostat equipped with both the heater and the cooling system.

Concentrations of O_2 in different solutions. The solubility of Oxygen also strongly depends on the presence of solutes.

Table 8.1 Concentrations of Oxygen in solutions containing sucrose or KCl.

1 ml. of a Solvent At 25°C	[O_2] nanomol	Nanogram Atom O (nmol O_2) x 2
H_2O	245	490
130 мM KCl	240	480
Sucrose 20 мM	241.8	483.6
Sucrose 200 мM	209.4	418.7
At (37°C) H_2O	217	434
130 мM KCl	212.5	425

Thus, a chamber with the volume of 0.65 ml contains, when filled with the KCl-based incubation medium, 156 nmol O_2 (312 ngA

O) at 25⁰C. These numbers for oxygen have to be used during calculations of the respiratory rates.

Traditionally, the mitochondrial respiratory rates were expressed in nanogram atoms O per minute per mg of mitochondrial protein. This was because it was thought that electrons react not with molecular oxygen but with the atom of oxygen to produce a molecule of water. On the other hand, mitochondria consume molecular oxygen; therefore it is also correct to express the respiratory rates as nanomol O_2 per minute per mg of mitochondrial protein.

Both methods are correct. It is simply necessary to remember about the conversion factor of 2 when comparing the published respiratory rates because the authors may use different methods of calculation.

Types of electrodes and monitor systems for measuring dissolved oxygen. There are several types of commercially available incubation chambers, Clark-type electrodes and oxygen monitors, which I will briefly discuss from the experimentator's point of view. I will describe only those systems, which I and my collaborators were working with.

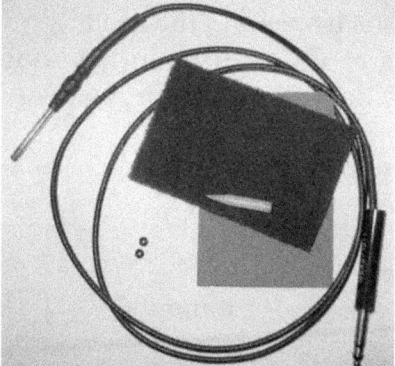

Figure 8.4. Combination Clark electrode with YSI cable. The electrode is shown together with some accessories for maintenance.

For 20 years of my work in the USA, I used the oxygraph and the clrak-type electrodes manufactured by the former Yellow Springs Instrument Company (YSI), which was formed in 1948. After Leland C. Clark, Jr. (1918-2005) invented his famous oxygen sensor, the YSI begun production of the Clark electrode, which acquired the name of the Standard YSI electrode (Fig. 8.4). This electrode was for several decades very popular among mitochondriologists. However, in early 2000 something went wrong during manufacturing of the electrodes because they become very unreliable – some of them would stop working after just few days of usage. The usual symptom of the malfunction of the electrode was that when being immersed into a KCl solution, the monitor showed 1.0, probably due to the short circuit between the platinum and the silver through microcracks.

After having so many troubles with commercial incubation chambers, I made a chamber, which I desined having in mind the possibility of simultaneous measurements of respiration and membrane potential. In addition, the latest commercial oxygen sensors became so expensive and often difficult to maintain. Therefore, I found it is easier, more reliable and much cheaper to make platinum electrode and chamber myself or order to a workshop.

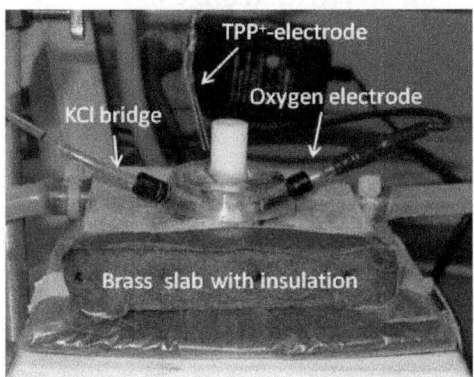

Figure 8.5. The custom made plastic chamber (0.65 ml). The chamber is shown with the standard Clark (right) and the custom made TPP-sensitive (rear) electrodes and agar/KCl bridge (left) connecting to Ag/AgCl reference electrode.

Figures 8.6 and 8.7 show a Microcathode and Mitocell of the Strathkelving oxymeter. This instrument with 200 μl chamber allowed me to spare mitochondria, which I isolated from the brains and spinal cords of very expensive transgenic SOD1 rats ($100 each). However, the software was very bad: I was not able to obtain the real time pictures of mitochondrial respiration as shown in the Fig. 6.1.

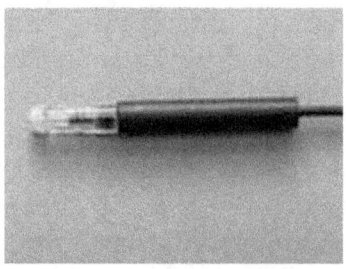

Figure 8.6. Microcathode Oxygen Electrode Model 1302.

This is a Clark-type polarographic electrode with a 22 micron diameter platinum cathode and silver/silver chloride anode connected by a buffered potassium chloride electrolyte solution. The electrodes are not temperature compensated and require controlled temperature environments (within 0.1°C).

Figure 8.7. Mitocell Miniature Respirometer Model MT 200 and MT200A.

This Mitocell Miniature Respirometer has a chamber volume of only 50-100 μl and was introduced for measurements on mitochondria isolated from biopsy samples. It can be used in any situation where sample size is limited. The base section contains an integral solid state, fixed speed magnetic stirrer. The 1302 microcathode electrode (BS4 69-3006) is inserted from beneath of the unit, and its projecting tip forms the base of the respirometer's chamber. The glass chamber unit is surrounded by a water jacket through which constant temperature water is circulated.

The Hansatech Instruments produces an oxygraph, which includes the oxygen electrode disc with a Clark type polarographic sensor, comprising a resin bonded central platinum cathode and a concentric silver anode (Fig. 8.8A). During measurements, the anode and cathode are linked by a 50% saturated KCl electrolyte solution. On top of the sensor disk is mounted the chamber, which has a variable volume from 1 ml to 0.2 ml. The chamber has a water circulating jacket for temperature control (Fig. 8.8B). The electrode disc is connected to a control unit, which supplies a polarizing voltage of 700 mV.

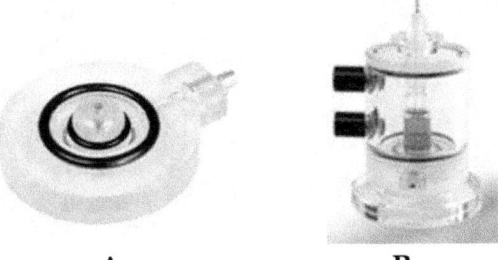

A   **B**

Figure 8.8. The Hansatech Instruments manufactures the oxygen measurement system.

The Harvard Apparatus presents the Batch/Flow Chamber System for Harvard/Instech Oxygen Monitoring System (Fig. 8.9A), which has two chambers (Fig. 8.9B) with the standard electrodes shown on Fig. 8.4.

As I have already mentioned, the currently manufactured standard electrodes are unreliable and expensive (two electrodes for the System were at $700.00 in 2012). The 600 µl chamber is also inconvenient: the glass cover, which is attached to the chamber with vacuum grease, often fells off during addition of reactants or mitochondria in the course of an experiment.

The drawbacks of all commercial Clark electrode chambers are: they do not allow simultaneous measurement of membrane potential and are very expensive.

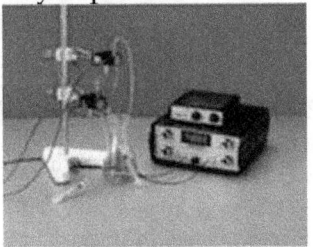 

Figure 8.9A   **Figure 8.9B.**

The Austrian Oroboros Company manufactures oxygraph, which allows simultaneous measurements of respiration and membrane potential, but it is even more expensive than other oxygen monitoring systems and is impractical to study isolated mitochondria.

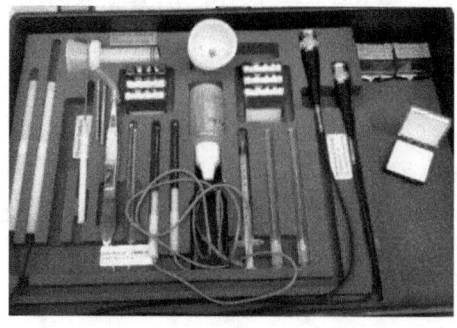

Figure 8.10. The Oroboros oxygraph (Austria). The instrument has two chambers of 2.0 ml volume (Left), and the kit for measurements mitochondrial membrane potential with the TPP$^+$-sensitive electrode (Right).

Although the Oroboros oxygraph does allow simultaneous measurements of oxygen consumption and estimation of membrane potential, the instrument is not worth the money it cost – around $40,000.00 (in 2011). The major drawbacks are the large volume (2 ml) of the chambers, and the TPP-sensitive electrode is too expensive and difficult to assemble and handle.

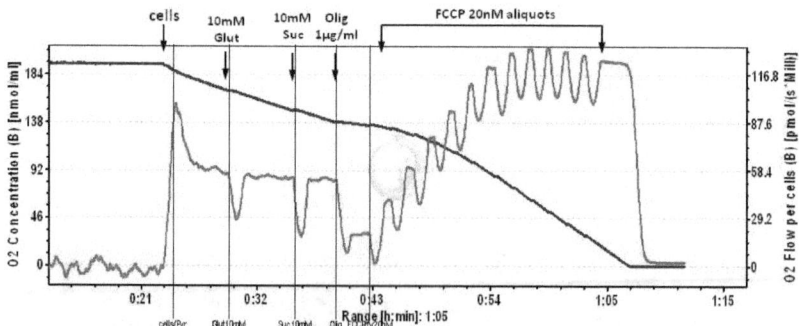

Figure 8.11. Registration of O2 concentration (blue line) and O2 flux (red line) in experiment with PC3 cells (2 x 106 cells/ml). Incubation conditions: 120 mM NaCl , 10mM KCl, 2mM KH2PO4, 2mM MgCl$_2$, 20mM MOPS 0.7mM CaCl$_2$, 1mM EGTA, pH 7.4, 10mM Pyruvate was added before cells. PC3 prostate cancer cells 2 x 10^6/ml. The figure was kindly presented by Dr. Z. Orynbaeva, Drexel University.

The software (Datalab), besides registering the rate of oxygen consumption, also shows the so called O_2 fluxes calculated for 1×10^6 cells (see Fig. 8.11).

Because of the very large chamber volume, the Oroboros instrument is poorly suited for characterization of mitochondria isolated from small tissue samples and very expensive transgenic animals. However, Oroboros instrument, in principle, can be a good contemporary substitution of the Warburg apparatus because it allows following respiratory activities of cells and tissue homogenates in prolonged experiments, which may be necessary for the metabolic studies and the effects of biologically active compounds.

Description of a custom made respiratory chamber for simultaneous measurements of respiration and membrane potential with a TPP$^+$-sensitive electrode.

For many years I used my custom made chamber, which was very inexpensive, reliable and convenient to use (Figs. 8.5. and 8.12). It was adapted for the commercial standard Clark minielectrode, a custom made TPP electrode (see below) and agar-KCl connective bridge to the AgCl reference electrode. As the commercial standard electrodes become unreliable and too expensive to replace on the regular basis, I have made a very simple platinum electrode, which was a reliable replacement for the commercial combination Clark electrodes (Fig. 6.5.13). The electrodes work well with the Instech Model 5300 Oxygen Monitor and other types of oxygen monitors, which accept the standard YSI cable (Fig. 6.5.4). The chamber shown on Fig. 6.5.12 was made of the transparent plastic with the detachable bottom, which was fixed with four small screws. Before fixing to the chamber, the surface of the bottom plate was smeared with vacuum grease to prevent wetting.

Instead of the relatively expensive hollow water jacket for water circulation surrounding the incubation chamber, I used a 15 mm thick copper or aluminum plate with a hole in the middle to accommodate the chamber. At one side, the plate was drilled through and two metal tips for connecting the tubes from the water circulating thermostat. The plate was insulated with a thin layer of foam plastic. This chamber holder is simple and inexpensive but perfectly efficient (Fig. 8.12.).

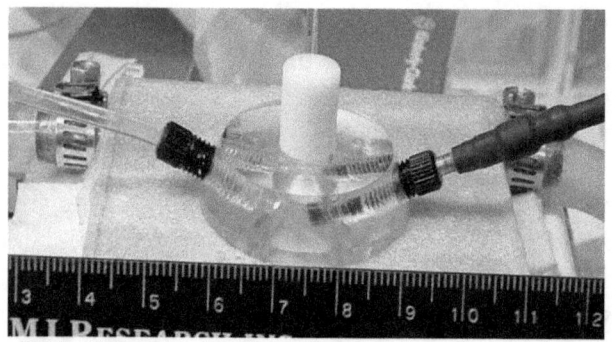

Figure 8.12 Custom made plastic chamber (0.65 ml) with standard Clark (right), custom made TPP-sensitive (rear) electrodes and agar/KCl bridge (left) connecting to the Ag/AgCl reference electrode.

The custom made platinum electrode was inserted into a Teflon cover with two holes, one for the platinum wire, the other to add mitochondria and chemicals to the chamber. This new cover (see Fig. 8.13 replaced the white stopper shown on Fig. 8.12. The hole in the chamber's body, which was designed for the standard Clark electrode, in this case served for connecting with the second Ag/AgCl reference electrode (Fig. 8.12) similar to the TPP$^+$ electrode.

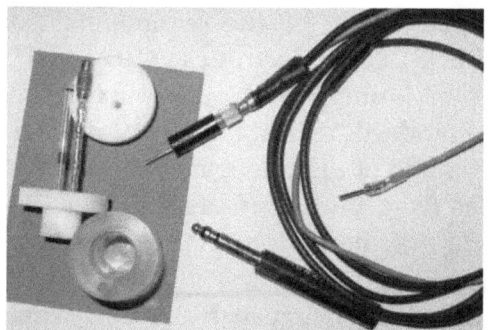

Figure 8.13. Custom made platinum electrode and chamber for measuring oxygen consumption.

Importantly, that with the custom made platinum and the custom made TPP$^+$ electrodes you have to have two separate Ag/AgCl reference electrodes. Do not use the single Ag/AgCl electrode connected to the chamber with two agar-agar/KCl bridges. In this case the electrodes will affect each other.

Preparation of the simple platinum electrode

1. To prepare the simplest Clark electrode, use the platinum wire with a diameter of about 1 mm. The working tip must be polished with a stone and cloth. Before soldering the platinum to the cable's connector, heat the platinum wire until red and deep into concentrated HNO_3, then wash thoroughly with water. This procedure will remove many contaminants from the platinum.

2. Wear rubber or plastic gloves to prevent contamination of the platinum with fat. Wash the platinum electrode with terahydrofuran, CCl_4 or ether.

3. Depolarize the electrode for 2 minutes at 2 V in 0.5N H_2SO_4 (1.4 ml of concentrated analytical grade sulfuric acid/100 ml H_2O. Do not forget that sulfuric acid MUST be added to water, not water to the acid), and change the polarity every 20 seconds. Finish with minus polarity (-) on the working platinum electrode. As a reference, also use the platinum wire.

4. The connecting wire to the oxygraph must be of pure copper or thin platinum wire and soldered to the non-polished tip of the platinum electrode with a tin, which contains no trace of silver. The simplest way is to use the standard YSI cable from the nonworking standard electrode. In this case solder the platinum to the central tip of the cable's connector taken from the disassembled old Clark electrode. To the outer gold-covered part of the cable solder the silver coated thin wire, which will be connected to the Ag/AgCl reference electrode. In this case the male tip of the connector is soldered to the cable, and the female tip is soldered to the reference electrode (Figure. 8.14).

5. Insert the platinum electrode into the Teflon cover of the chamber in a way that the polished tip of the platinum is just above the inner surface of the cover. Apply 1-2 drops of 5% solution of polysterol in terahydrofuran or CCl_4. Be careful, there has to be no bubbles around the platinum and the procedure must be carried out under hood with good ventilation. Both CCl_4 and Tetrahydrofuran vapors are very toxic for the liver.

6. Dry the electrode at 37° C for several hours, and soak for a night in water.

This electrode will reliably work for many days or weeks. Store electrode dry, but once in a while clean the polysterol cover with

Tetrahydrofuran, then depolarize the electrode as shown in the step 3 and renew the polysterol cover.

As a reference electrode it is best to use the commercial Ag/AgCl electrode shown in Figures. 8.14. If you do not have one, you can make the reference electrode as described below.

IMPORTANT. Most chambers made of transparent plastic have tiny cracks, which you do not see. However, if you will wash the chamber with 70% ethanol, these cracks will enlarge and destroy the chamber. Therefore, do not wash the chamber with pure ethanol, but you can use 40-50% ethanol. It is better to avoid ethanol completely. If you use uncouplers, such as CCCP or FCCP, washing the chamber with ample amount of water will remove uncouplers completely. Inhibitors, such as rotenone, stick to the plastic so tightly that no ethanol washing will remove the inhibitor completely. For experiments with rotenone I usually had a separate chamber. However, if you had to use rotenone, antimycin or oligomycin, I recommend after washing the chamber with water, fill the chamber with incubation medium, add substrate and 20-30 mg of mouse or rat liver mitochondria and incubate for 3-5 minutes. Repeat this procedure 2-3 times and your chamber will have no trace of inhibitors. It is easy to isolate liver mitochondria for this purpose and for testing your equipment. This will save you precious mitochondria from other organs.

Preparation of the Ag/AgCl Reference Electrode
1. A sheet of metallic silver of 99.99% purity clean with a glass powder or very thin sand paper.

2. Solder the connective wire coated with silver, and use soldering tin also containing silver. You may cover the soldering with protective paint or polish to protect from corrosive effect of KCl.

3. Wash the silver thoroughly with water.

4. Dip the sheet of cleaned silver into 10% HNO_3 solution for 2 minutes. There must appear bubbles and the surface turns black. From this time, keep the electrode from exposure to the bright light. Wash thoroughly with water

5. Dip into the concentrated ammoniac NH_4OH for 10 minutes.

6. Wash in water for 30 minutes (keep in the dark)

7. For 12 hr. (for a night) dip the electrode into 3M KCl solution and apply (+) to the silver electrode and (-) to platinum wire (not the electrode) 0.65V (you can use the polarograph as a source of the charges). You can also use a 2 V alkaline battery for the purpose.

8. The Ag/AgCl electrode must be assembled into the black plastic or dark glass bottle. In the plastic cover two tight holes are made, one for the connective wire, the other for the agar-agar-KCl connective bridge. The wire should be made of copper coated with silver or silver wire, which is soldered to the silver plate with the tin containing silver (the tin without silver also acceptable). The bottle containing Ag/AgCl electrode is filled with the 3 M KCl.

The commercial reference electrodes (Fig. 8.14) are very convenient and reliable. They can be placed into a 50 ml centrifugal plastic tube filled with 3 M KCl with two holes in the cover (Fig. 8.14). Because KCl can "creep" along the tubes and the vessel's walls, and cause short-circuits, the assembled reference electrode and the tubing require constant attention and cleaning.

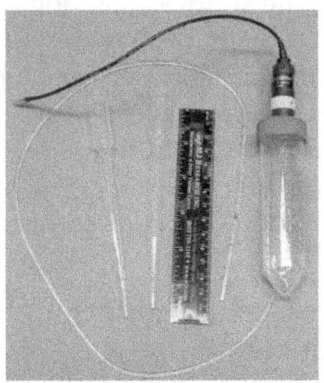

Figure 8.14. Commercial reference Ag/AgCl electrode and Nalgene tube as agar-agar-KCl bridge.

The reference electrodes must be connected to the incubation chamber with the low resistance connection bridge, which is described below. The end glass tips filled with the KCl-agar-agar are much better than the porcelain connective tips because they have much lower resistance, are reliable and cheep. The porcelain tips are expensive, easily get stuck with protein, and thus increase resistance and noise. This happens very quickly if the incubation medium contains Tris buffer. Tris forms with AgCl insoluble compound, which obstruct the porcelain filter.

Preparation of the KCl-agar-agar connective bridge

This simple and inexpensive device provides very efficient connection between the reference electrode and the experimental chamber. The KCl-agar-agar bridge has low electrical resistance and therefore measurements of membrane potential or pH changes become much less affected by noises, as compared with a connective KCl bridge when one end of the plastic tube filled with KCl is stopped with a porcelain filter.

In a small glass bottle with tight cover dilute 1.5 g KCl in 5 ml of water (or take 5 ml of 3M KCl solution) and add 150 mg of agar-agar. The tightly covered bottle heat in the boiling water bath until the solution becomes completely transparent and even. Do not shake the bottle while heating. Using a PVC Tygon capillary tube, fill the glass capillary tubes 20-30 mm long with the hot KCl-agar-agar. To prevent sliding out of the KCl-agar-agar gel out of the glass tip, the outer end of the glass tube should be narrowed in the gas flame, or bend the capillary tube in a "Z" shape. The simplest solution is to connect two 20 mm long glass capillary tubes with 20-25 mm long plastic capillary tube and fill with the hot KCl-agar-agar gel, leaving one glass tube half empty (Fig. 8.14). Connect the glass tubes with a thin transparent PVC Tygon tube filled with 3 M KCl solution. For this, use a small syringe and a long needle. Such a bridge can work for months. Store the glass tips filled in advance with KCl-agar-agar in the 50 ml plastic tube filled with 3 M KCl.

Simultaneous measurements of mitochondrial respiration and membrane potential

In the section "General considerations", we discussed the metabolic states of mitochondria and how respiratory activities depend on the type of a substrate and substrate mixtures. In this section, we will discuss how mitochondrial respiration and membrane potential, which are the most important functions for understanding mitochondria, depend on the isolation method and how simultaneous measurements of these two parameters allows resolving experimental puzzles.

Historically, it happened that most of our current knowledge about mitochondrial functions was obtained by studying rat or mouse liver mitochondria. The activity of oxidative phosphorylation in liver mitochondria strongly depends on the metabolic state of the animal, but relatively little on the compositions of the isolation and incubation media.

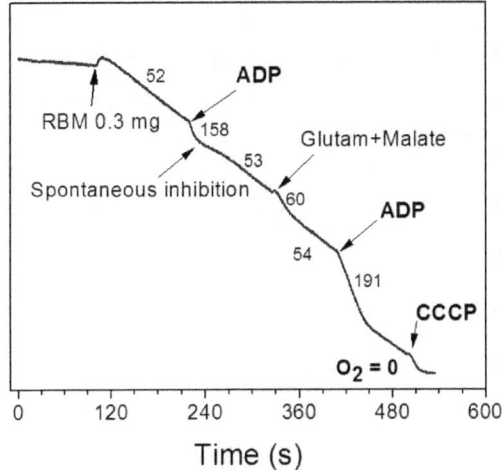

Time (s)

Figure 8.15. Respiratory activity of rat brain mitochondria isolated in the absence of defatted BSA and oxidizing succinate. Experimental conditions. Rat (Sprague Dawley) brain mitochondria were isolated in the absence of 0.1% defatted BSA. "Full" medium, succinate 5 mM. Additions: RBM 0.3 mg, ADP 150 μM, CCCP 0.4 μM, glutamate 5 mM, malate 2 mM. The numbers at the trace are the rates of O_2 consumption in nmol/min/mg protein RLM.

Quite different situation is observed with mitochondria from other organs. The respiratory activity of brain mitochondria, for example, depends little on the metabolic state of experimental animal, but is highly dependent on the composition of the isolation and incubation media. In addition, unlike liver mitochondria, mitochondria from brain, heart, kidney and other tissues are commonly isolated in the presence of defatted BSA in order to protect mitochondria. In the liver mitochondria isolated from the overnight starved animals, mitochondria oxidize well glutamate + malate and succinate. In the presence of rotenone mitochondria do not accumulate oxaloacetate and thus do not develop inhibition of respiration in State 3 and State 3U. For this reason, succinate + rotenone became kind of "standard" substrate for measuring respiratory activity of mitochondria regardless of their origin. We have already discussed that usage of rotenone as a "standard" additive to study normal mitochondria is a nonsense.

It is instructive to study what will happen if we isolate brain mitochondria without the presence of BSA in the isolation medium, and provide the isolated brain mitochondria with succinate without the presence of rotenone? The results are presented in Figure 6.5.15.

Figure 8.15 shows that addition of ADP caused a rapid inhibition of respiration by accumulating oxaloacetate (OAA). In the presence of uncoupler the inhibition of respiration developed even faster and was more pronounced (not shown). Addition of glutamate + malate removed OAA in the transaminase reaction, and thus released inhibition of respiration. The interpretation of the experiment presented in Figure 8.15, when based only on the polarographic trace, was not a simple task.

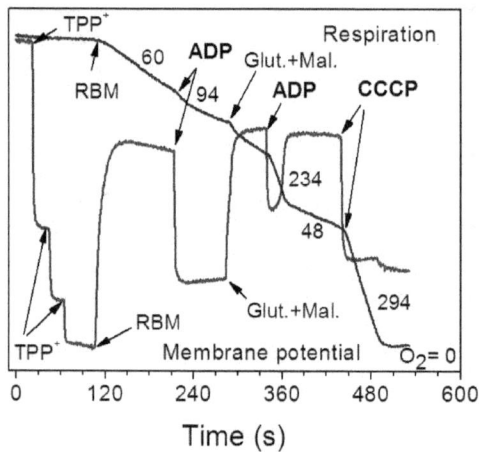

Figure 8.16. Respiratory activity and membrane potential of rat brain mitochondria isolated in the absence of defatted BSA. Icubation conditions as in Fig. 8.15.

Figure 8.16 presents similar experiment to the one shown in Fig. 8.15, but this time, both respiratory rates and membrane potential were recorded simultaneously. The reader can see how much more important information we have now at hand, and how much more reliable becomes our interpretation. Our conclusion was that brain mitochondria isolated from the rat's brain in the absence of BSA were not uncoupled by fatty acids (this is a common belief about protective effect of BSA on mitochondria during isolation procedure), but low oxidation of succinate was caused by specific inhibition of SDH by OAA. That mitochondria were not damaged was proved by providing RBM with a mixture substrates that prevent inhibition of SDH (Fig. 8.17).

The simultaneous measurements of respiration and membrane potential become of particular importance when we encounter mitochondrial dysfunctions caused by mutations in mtDNA, poisoning with toxins, and when we encounter variations in mitochondrial metabolic phenotype. In this mode of work, both the platinum and the TPP+ electrodes were connected correspondingly to the oxygen monitor and the pH-meter, which functions in the mV mode. The instruments were connected to the two-channel recorder, and via the two-channel amplifier (I used one from Warner Instruments) to the data acquisition system.

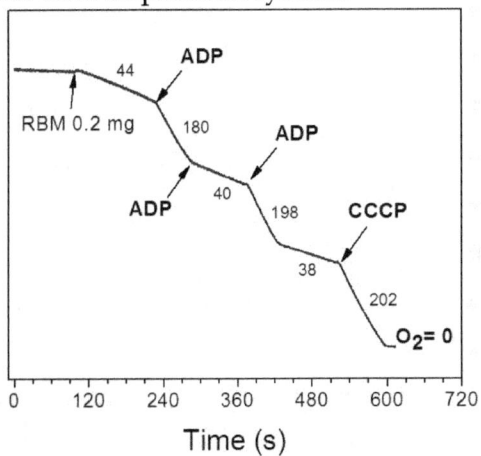

Time (s)

Figure 8.17. Respiratory activity of brain mitochondria isolated in the absence of defatted BSA. Experimental conditions as in Fig. 8.15. The same RBM 0.3 mg oxidized succinate 5 mM + glutamate 5 mM + pyruvate 2.5 mM + malate 2 mM. For details see Panov et al., 2009, 2010.

Simultaneous measurements of respiration and membrane potential not only save time and mitochondria, but give important information allowing for better interpretation of the experimental data, particularly if the results were not typical. Let us consider several examples.

Figure 8.18 illustrates the relationships between changes in the respiration rate and membrane potential during oxidation by isolated rat heart mitochondria (RHM) L-palmitoyl-carnitine, which is regarded as a physiological substrate for the heart. Generally, an increase in respiration, caused by energy utilization (oxidative phosphorylation, Ca^{2+} transport, etc), is accompanied by a decline in membrane potential. However, sometimes the straightforward relations between these two parameters become complicated and difficult for interpretation.

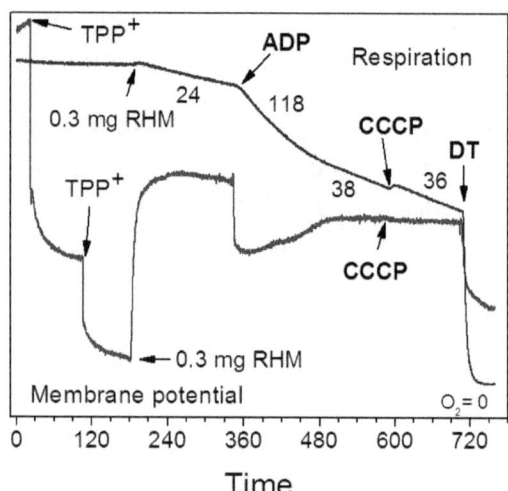

Time

Figure 8.18. Respiratory activity of rat heart mitochondria oxidizing L-palmitoyl-carnitine 50 μM + malate 2 mM. Experimental conditions as in Fig. 8.15.

In this example, the absence of the membrane potential collapse upon addition of 400 nM (or 400 pmol/ml CCCP) was caused by binding of the uncoupler to palmitoyl-carnitine and ethanol (which was a solute for palmitoyl-carnitine). In another experiment (not shown), a 3-fold higher concentration of CCCP did cause collapse of the membrane potential, but still failed to stimulate respiration because transport of palmitoyl-carnitine into matrix is energy-dependent. This is a good example of how the knowledge of the simultaneous responses of respiration and membrane potential may help to solve an unknown problem during the experiment.

There are many other situations when the results on oxygen consumption or membrane potential registered separately would be difficult or impossible to interpret. As was shown in Figures 8.15 and 8.16, brain, as wells as spinal cord mitochondria (Panov et al. 2010), are highly sensitive to inhibition of succinate dehydrogenase by oxaloacetate. The degree of SDH inhibition depends not only on the isolation conditions, but varies strongly between the animal species.

Figures 8.19A and 8.19B illustrate experiments with brain mitochondria from FVB mice (MBM), which displayed a phenotype with very high intrinsic inhibition of SDH by endogenous oxaloacetate. Fig. 8.19A shows that in MBM isolated with BSA, the State 4 succinate oxidation was accompanied by increase in the membrane potential. However, it took about 2 minutes for the membrane potential to reach the maximum level. Compare with figure 8.16, where RBM from Sprague Dawley rat isolated without BSA were energized within 20 sec.

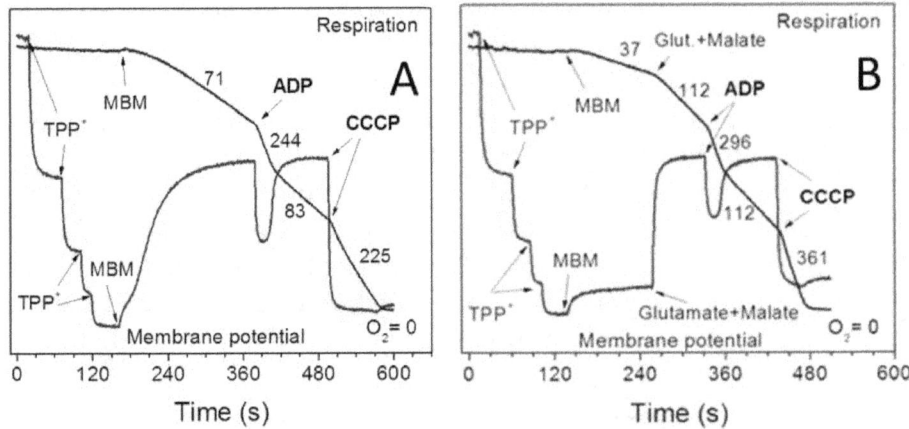

Figure 8.19. Oxygen consumption and membrane potential during oxidation of succinate by mouse brain mitochondria isolated in the presence of BSA. A. Mouse (FVB) brain mitochondria were isolated in the presence of 0.1% BSA. B. Mouse (FVB) brain mitochondria were isolated without BSA. Additions: Mouse brain mitochondria 0.3 mg, TPP+ 0.5 µM (Final [TPP+] 2 µM), ADP 150 µM, CCCP 0.4 µM, glutamate 10 mM + malate 2 mM.

When FVB mouse brain mitochondria were isolated without BSA (see Fig. 8.19B), the rate of the State 4 respiration was so low that mitochondria were unable to generate membrane potential. Addition of glutamate + malate instantly released the inhibition of SDH by removing oxaloacetate in the transaminase reaction (Fig. 6.5.20). The effects of BSA on oxidation of succinate by brain mitochondria were discussed in Panov et al. (2010).

The mechanisms, which are responsible for the differences in metabolic phenotypes at the level of succinate dehydrogenase remain largely unstudied. These might be differences in the affinity of SDH to oxaloacetate and/or succinate, and so on.

Because the activity of succinate oxidation in State 4 and the associated reverse electron transport are one of the major determinants of the superoxide radical generation by mitochondria, the variations in mitochondrial metabolic phenotype may greatly affect susceptibility of animal species to pathological agents in animal models of diseases. An example of influence of a change in the mitochondrial metabolic phenotype was described in a paper on the SOD1 transgenic model of Amyotrophic Lateral Sclerosis (Panov et al. 2012). In this paper we provided evidence that a change in the metabolic phenotype of neuronal mitochondria, which evidently occurred in the stock animals between 2005 and 2008, resulted in the loss of the morbidity in the transgenic Sprague Dawley rats, even though they expressed the human mutated SOD1 gene (Panov et al. 2012. In: Amyotrophic Lateral Sclerosis. Ed. M.H. Maurer. Chapter 9, p-225-248. Intechweb.org free download).

In the rat heart mitochondria (RHM) from the Sprague Dawley rats, BSA had no effect on the intrinsic inhibition of SDH by endogenous oxaloacetate. It persisted regardless whether RHM were isolated with or without BSA. The inhibition of SDH by endogenous OAA was released metabolically by transamination of glutamate.

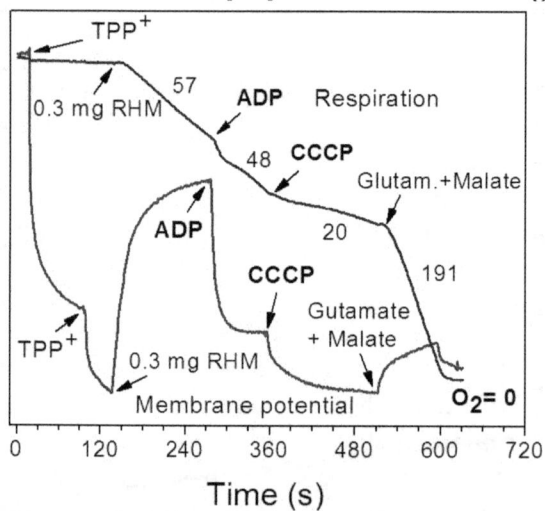

Time (s)

Figure 8.20. Oxidation of succinate by rat heart mitochondria oxidizing succinate 5 mM. Incubation conditions as in Fig. 8.16.

Figure 8.20 shows a typical experiment with RHM isolated in the medium without BSA, which is identical to the experiment with RHM isolated in the presence of 0.1% BSA (not shown). The RHM did oxidize succinate in the State 4, and the membrane potential was high. However, upon addition of ADP, the mitochondria failed to maintain oxidative phosphorylation: respiration increased only briefly, and membrane potential dropped significantly. Addition of uncoupler

(CCCP) also failed to stimulate respiration. Glutamate + malate almost instantly released the inhibition of SDH by removing oxaloacetate from the enzyme.

These examples illustrate that interpretation of the polarographic data is not always simple. In addition, there are differences between mitochondria from various organs and tissues, and even for the same organ from animals of the same strain you may find significant changes in mitochondrial metabolism over the years. This is probably associated with the fact that the animal breeding companies may supply animals from different family nodes.

Thus, in order to characterize the mitochondrial metabolic phenotype, it is necessary to conduct full analysis of mitochondrial respiration using a set of different substrates and their mixtures. Every organ has a specific preference for substrates. For example, brain and spinal cord mitochondria show maximum rates of oxidative phosphorylation during simultaneous oxidation of glutamate + pyruvate + malate, or succinate + glutamate + pyruvate + malate. Even with this complex substrate mixture, in some animals malate inhibited production of ROS; in other animals, malate stimulated production of ROS. The research on the roles of metabolic mitochondrial phenotype in pathogenesis of diseases has just begun and has a great perspective in the Future.

It is clear, that the old ("classical") methodology to study mitochondrial respiration and other functions using only one type of a substrate is not sufficient, because it usually does not reflect the physiological metabolic phenotype of mitochondria. The wide usage of succinate + rotenone is completely unacceptable. Rotenone disrupts functional interactions between components of the respiratory chain, the tricarboxylic acid cycle enzymes, the redox states of pyridine nucleotides, and the regulatory effects of metabolites. Thus, with succinate + rotenone it becomes impossible to study many mitochondrial functions and regulatory mechanisms, such as the reverse electron transport and associated ROS production, intrinsic regulation of the succinate dehydrogenase activity and transaminase reactions. Without correctly investigating mitochondrial respiration we cannot interpret and understand other functions of the mitochondria: membrane potential, ion transport and production of ROS, permeability transition.

In some papers, the researchers provide only information about respiratory control ratio (RCR), stating that mitochondria were reasonably good because the RCR values were high. However, by itself the RCR is meaningless. This is because $10/2 = 5$ and $100/20 = 5$. Only, if we have information about the actual rates of oxygen consumption during oxidative phosphorylation (State 3) and the resting metabolic state (State 4), RCR has some meaning.

In the next sections I will describe how to calculate the rates of mitochondrial respiration from the polarographic chart, what meaning is behind each of the metabolic state, and what additional information can be obtained about the quality and the metabolic states of the mitochondria.

However, the first and most important condition for obtaining meaningful data on mitochondrial respiration and membrane potential is to perform the polarographic experiment correctly.

CHAPTER 9

How to correctly conduct the Polarographic Experiment.

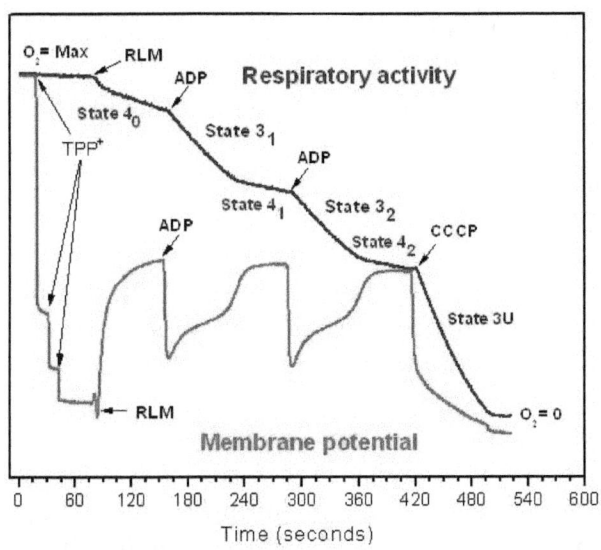

Figure 9.1. Polarographic and membrane potential traces obtained with rat liver mitochondria oxidizing glutamate 10 mM + malate 2 mM in various metabolic states. Additions: Rat (Lewis) liver mitochondria 0.5 mg, ADP 150 μM, CCCP 0.4 μM, TPP+ - 1st addition 10 μM, two others - 0.5 μM each (final [TPP+] = 2.0 μM).

Figure 9.1 will serve us as an example of a typical polarographic experiment. The figure shows changes in the rates of oxygen consumption and membrane potential of the rat liver mitochondria oxidizing glutamate + malate. Importantly, the animal starved overnight before the sacrifice. I have mentioned above that liver mitochondria are very sensitive to the type of a diet and the time passed after the last feeding. Therefore, most researchers, for the sake of better reproducibility of the results, take into experiments animals that starved overnight. I have selected this particular experiment also to illustrate how unusual can be the patterns of the membrane potential trace during oxidative phosphorylation. Though, in the Fig.

9.1 the membrane potential trace serves only as an illustration, and we will discuss the polarographic trace only.

The outcome of any experiment depends on how it was conducted. The polarographic experiments require several conditions that have to be satisfied in order to obtain correct results. Therefore, we begin with the discussion of experimental conditions and the sequence of actions during the experiment.

Experimental conditions and the sequence of the experimental steps

The incubation chamber has to be tightly closed with a stopper to prevent contact of the incubation medium with air. A narrow hole in the stopper allows to make additions of chemicals and mitochondria into the chamber.

The solubility of O_2 is highly sensitive to changes in the ambient temperature. Therefore, the temperature of the chamber has to be stabilized with a circulating thermostat. A simple and inexpensive design of the chamber holder for circulating thermostat was described above (Fig. 8.12).

In order to establish the magnitude of the temperature-dependent drifts of $[O_2]$, register the signal for 5-6 minutes, and then add 20-30 μl of the freshly made sodium dithionite ($Na_2S_2O_4$) solution (Dithionite) to reach the zero $[O_2]$ point. Normally, there should be no drift, or the drift should not exceed 2-3% of the whole scale per minute. The signal's scale is the distance between the points on the chart for the maximum and zero $[O_2]$. This procedure helps, if you do not have a chart recorder and have to calculate your results from the computer printed charts.

At the start of a new experiment, fill the chamber with the incubation medium, close with the stopper, and before you turn on the recording and add mitochondria, wait until the temperature and the drift will stabilize. Without a thermostat, this may take several minutes. This is why the circulating thermostat may save a lot of time during experiments.

An important issue is the time of addition of a substrate, or a mixture of substrates. If you do not have the task of establishing mitochondrial respiration on endogenous substrates – the substrates that are contained inside mitochondria, it is much more practical to add substrates into the incubation medium at the beginning of the experiment before you close the chamber. Liver mitochondria, as well mitochondria from the heart and skeletal muscle, contain enough substrates to maintain respiration and energization for several minutes. But they will be exhausted within several minutes. Mitochondria from the brain (BM) and spinal cord (SCM) do not contain endogenous substrates. Therefore, with BM and SCM added to the incubation medium without substrates, there will be no oxygen consumption at all. Because mitochondria in the working stock suspension will gradually consume the endogenous substrates, you have to study respiration on endogenous substrates using only the freshly isolated mitochondria.

In most experiments the respiratory substrates have to be added to the incubation medium before addition of mitochondria. It is practical to add substrates right after you filled the chamber with a known volume of the incubation medium. In this way it is easy to maintain the same concentrations of substrates in all experiments.

If you measure simultaneously respiration and membrane potential with a TPP$^+$ electrode, or use the membrane probe for oxygen (instead of a platinum electrode), you must know that ethanol, DMSO and other hydrophobic solvents affect the membrane-based sensors. Therefore, you have to add substrates, for example palmitoyl-carnitine that is dissolved in ethanol, or some chemicals or extracts from plants dissolved in ethanol or DMSO, before you add TPP$^+$ and mitochondria. You have to wait for several minutes to allow membrane sensors to accommodate to new conditions. In this way calibrated additions of TPP$^+$ will preserve their significance.

Commentary 1. In some oxygraphs, Strathkelvin, for example, calibration of the instrument is made using two points: water for maximum oxygen and zero oxygen in water in the presence of Dithionite. For maximum O_2 content, water must be saturated by intensive stirring for 15-20 minutes. This is necessary to do because researchers often take bidistilled water stored in closed glass containers, which might not be saturated with air oxygen.

Commentary 2. For preparation of the stock solutions of substrates, ADP, and some other stable chemicals (EGTA, EDTA, Mg^{2+}), prepare 50-100 ml of 20 mM MOPS and adjust pH to 7.2. This is physiological pH in the cell.

Commentary 3. The stock solutions of some substrates can be prepared in advance and stored at -20 for many months. It is convenient to make additions to incubation chamber or cuvette using the same pipetter. The 10 µl pipetter is convenient for most cases to add to 1 ml. Thus the stock solution should be 100 times more concentrated. Since some substrates are relatively strong acids, at high concentrations the stock solutions will be too acid, and you have to adjust pH to 7.2 using KOH. Therefore, it is much more practical to use K^+ or Na^+ salts of the substrates. In most cases, with 20 mM MOPS buffer as a solvent, the shift of the pH will be small (< 0.3 pH unit), which you can ignore (but you have to check pH anyway). I recommend the following concentrations of the substrates: (the first number is final concentration, the second number is the stock concentration) Succinate Na – 5/500 mM; Glutamate Na – 10/1000 mM; Malate Na – 2/200 mM; L-Palmitoyl-carnitine 25-50 µM/2.5-5 mM (dissolved in 50% ethanol). Using analytical scales you weigh the necessary amount, bring into a glass beaker and add 6-7 ml of 20 mM MOPS, add a magnetic stirring rod and dissolve the chemical. After the substrate will dissolve, bring the solution into a 15 ml plastic measuring tube and by adding drops of 20 mM MOPS bring the volume to 10 ml (leveling with low meniscus). Check pH, and if necessary adjust it to 7.2. After that, pipette 1 ml substrate aliquots into 1.5 ml plastic tubes with the cover. Mark the tubes accordingly. I found convenient to have tubes of different color for each substrate. Store substrates at -20°C. If you have opened a tube with the substrate, keep it in a refrigerator for daily use.

Commentary 4. Pyruvate, α-ketoglutarate and oxaloacetate are ketoacids, which are unstable. Therefore you have to prepare the stock on the day of experiment, and do not use the next day. If you use the sodium or potassium salts, you dissolve these substrates in the incubation medium and do not need to adjust pH. I used the following concentrations of these substrates: Pyruvate Na - 2.5/250 mM; α-ketoglutarate – 10 mM. In this case, it is easier to prepare 10 ml of the substrate solution in the incubation buffer. Oxaloacetate usually is not used as a substrate, but as an inhibitor of SDH.

Commentary 5. Do not buy DL-palmitoyl-carnitine. D-palmitoyl-carnitine inhibits oxidation of physiological L-palmitoyl-carnitine. Use only L-palmitoyl-carnitine as a substrate.

Little Trick. To make Life easier, I used the following trick. Say, you need to prepare 250 mM stock of Pyruvate for future experiments, and I do not want to do it every day. To prepare 1 ml of 250 mM stock of sodium pyruvate, Mol. Mass. 110. Step 1. Determine how much Na-pyruvate you need in mg/1 ml = (250 mM x 110)/1000 = 27.5 mg/1 ml; Step2. You plan to add 10 µl of the stock solution, so you do not need 1 ml of the stock. What, if you will weigh only 7 mg of Na-pyruvate then how much buffer you have to add? Solve the proportion: 27.5 mg/1 ml = 7 mg/X ml; X = 7/27.5 = 0.255 ml or 255 μl. Therefore, you make 5-10 weights of random mg from 5 to 8 (any), mark the tubes, and store in refrigerator for the future. On the day of experiment I take any tube, solve the proportion and have the working stock ready.

Addition of mitochondria. After the signal at the maximum [O_2] has stabilized, add mitochondria into the chamber. Mitochondria in the working suspensions very quickly precipitate to the bottom. Therefore, do not forget prior to addition to vortex thoroughly the suspension of mitochondria. The optimal amount of added mitochondria will depend on the chamber volume and the type of mitochondria. For the less active rat liver mitochondria, the optimal concentration of the mitochondrial protein is about 0.5 mg/ml. For the much more active brain, heart and kidney mitochondria, the optimum concentrations are 0.3-0.2 mg/ml. Remember, that in general, mitochondria from mice have by 30-40% more active respiration than the corresponding rat mitochondria.

If you measure simultaneously membrane potential, TPP+ is added before mitochondria in two or three aliquots. The added mitochondria should not consume all TPP+ from the incubation medium. After the maximal level of the membrane potential in State 4 was achieved (that corresponds to the maximum removal of TPP+ from the medium), the signal of the remaining TPP+ in the medium should be at least at 1/3 or 1/2 of the signal of the first addition of TPP+ (see the red line in Fig. 9.1 after addition of RLM). If you need to use a higher concentration of mitochondria, add more TPP+. Do not use TPP+ at final concentrations higher than 2.5 – 3.0 μM.

After addition of mitochondria to the incubation chamber, mitochondrial respiration is in the metabolic State 4, or resting respiration. In this metabolic state, mitochondria do not perform any significant work, and they supposedly have to achieve the maximal level of energization, and respiration has to be at minimum. For convenience, I designate this metabolic state as the State 4_0, in order to stress that this is the initial metabolic state before addition of ADP. It is important to remember that at State 4_0 there are no adenine nucleotides outside mitochondria. The significance of this will be discussed in Chapter 10.

Upon addition of mitochondria into the chamber, the O_2 trace may shift slightly to the left. This will happen if the volume of the added suspension of mitochondria was large (the suspension was not concentrated enough), and because the temperature of the added suspension was lower than in the chamber. If the chamber's temperature was maintained by a circulating thermostat, then the signal may be shifting to the right (towards less oxygen) rather rapidly. This is because in the concentrated working suspension the mitochondria are anaerobic and therefore will rapidly consume oxygen to restore the mitochondrial pool of exchangeable adenine nucleotides. In either case, the signal will stabilize and after some time achieve a steady state.

Mitochondria contain a rather constant amount of adenine nucleotides (AN) – ATP, ADP and AMP. ATP and ADP can be exchanged with the extramitochondrial adenine nucleotides via the specific carrier - adenine nucleotide translocase (ANT), while AMP cannot be transported across the inner membrane. Therefore, ATP + ADP are designated as the "exchangeable pool of AN". In a concentrated suspension, mitochondria are close to anaerobiosis, and therefore mitochondrial pool of AN usually contains very little or no ATP, much of ADP, and a significant portion of AMP. A particularly large amount of AMP may mitochondria from the liver of animals at the period of absorbtion, and heart mitochondria from hypoxic animals. The pool of exchangeable AN will be restored during incubation in the State 4_0. Using HPLC, we have established that with rat liver mitochondria it takes about 2 minutes to convert mitochondrial AMP to ADP and ATP. Therefore, after addition of mitochondria, allow the mitochondria to respire in State 4_0 for at least 2 minutes. With glutamate + malate or α-ketoglutarate as substrates, you can shorten this time to 1.5 or 1 min. Glutamate (after transamination to α-ketoglutarate) and α-ketoglutarate activate substrate phosphorylation inside mitochondria, and thus facilitate the conversion of AMP to ADP and ATP. But it is more practical to have the initial State 4_0 respiration for 2 minutes. In this case the rates of the State 3 respiration will be much more reproducible.

Addition of ADP. After 2 minutes of incubation in State 4_0, you add to the chamber ADP. Mitochondria begin to phosphorylate ADP to ATP. This is oxidative phosphorylation, or metabolic State 3 (which we designate as State 3_1). Phosphorylation of ADP consumes energy;

therefore membrane potential will temporarily drop to a certain level (usually by 30-40 mV).

Question: How much add ADP? This you can determine using the following criteria: **1.** Respiration in State 3 has to be long enough to allow reliable determination of the oxidative phosphorylation rate. **2.** Usually, you want also to determine in the same experiment the rate of uncoupled respiration, and probably a response to the 2d addition of ADP, or some other chemical. Therefore, the State 3_1 should not be too long. In practice, a reasonable final concentration of added ADP is between 150-250 μM. **3.** The type of the substrate will also affect the reasonable concentration of ADP: with glutamate or pyruvate, mitochondria have "three phosphorylation sites" (acttually thre proton pumps), whereas with succinate – "two phosphorylation sites". Therefore, with succinate it will take longer and more oxygen for mitochondria to convert ADP to ATP. **4.** If mitochondria are damaged or slightly uncoupled, it will take more time and oxygen to phosphorylate the same amount of ADP.

Addition of 150 μM ADP with 0.5 mg liver mitochondria or 0.3 mg brain/heart/kidney mitochondria is optimal for most experiments.

Commentary 6. ADP solution, which you prepare using incubation buffer, is stable and you can prepare stock aliquots of 0.5 ml into tubes and store at -20. ADP remains stable when frozen. If you defrost a tube, use it or throw away. You can use the same tube until it will finish, store after work in refrigerator but do not freeze. I used 10 μl additions into the chamber calculating the stock based on the chamber's volume. Say, my chamber has volume of 0.65 ml. So, dilution is 65. To obtain 150 μM as final concentration in cuvette, the stock solution has to be 0.15 x 65 = 9.75 mM.

Commentary 7. Some researchers use very high concentratios of ADP. This is not wise, because mitochondrial have in the intermembrane space high activity of adenilate kinase. Th enzyme has relatively low affinity for ADP, but at concentrations at about 0.5 mM and higher it will convert ADP into ATP + AMP.

After mitochondria finish phosphorylation of added ADP, the mitochondria return back to energized state: membrane potential returns back (in ideal) to the previous high level and respiration slows down. This is once again the State 4, which we designate as the State 4_1. Allow this State 4_1 to last at least for 1 minute before adding either new ADP or uncoupler, and the overall picture of the experiment will look nice (see Fig. 9.1). Keep in mind that State 4_1 is fundamentally different from the State 4_0 by the presence of ATP and ADP outside of the mitochondria.

Addition of Uncoupler. If you make the second and the third additions of ADP, you will repeat situations at the States 4 and State 3, we just discussed. Correspondingly, the metabolic states will be designated as State 3_2, State 4_2, etc.

However, if you add an uncoupler, the situation will be quite different. Uncouplers are chemicals, which protonate the inner mitochondrial membrane by carrying protons across the membrane. This will collapse the transmembrane ΔpH, mitochondria become deenergized, and the membrane potential ($\Delta\Psi$) also collapses (see Fig. 9.1). These events release the flow of electrons down the respiratory chain from the control by mitochondrial energization, and the respiration should increase to maximum (See Fig. 9.1). This is the metabolic state, which is designated as State 3U, or uncoupled respiration. This ideal response to uncoupler classically occurs in mitochondria oxidizing succinate + rotenone, a non-physiological situation. In Real Life, mitochondrial respiration far too often does not respond by increased oxygen consumption upon addition of an uncoupler. We will discuss deviations from the "classical response" of uncoupled respiration in Chapter 10.

Question: What are uncouplers, how to choose one, and how much uncoupler to add?

Uncouplers are hydrophobic weak acids that can cross the inner membrane in a protonated form. Since pH in the matrix is more alkaline, the acid molecules release protons and thus diminish ΔpH. The negatively charged anions move electrogenically to the positively charged outer surface of the inner membrane and acquire new protons. This process is repeated many times, and thus few molecules of a protonophore can efficiently dissipate ΔpH. In fact, the hydrophobic weak acids work as carriers for protons, and therefore are called protonophores. Because protonophores release mitochondrial respiration from the control by energization, they are also called uncouplers.

The first known uncoupler used by researchers was 2,4-dinitrophenol (DNP).

2,4-Dinitrophenol is a yellow, crystalline solid that has a sweet, musty odor. It is soluble in ethanol and other hydrophobic solvents. However, the uncoupling capacity of DNP is relatively weak, and requires too high doses (> 10 μM) to uncouple *in vitro* the actively respiring well coupled brain or heart mitochondria. For this reason DNP is not currently used as an uncoupler with the isolated mitochondria.

The most commonly used uncouplers are CCCP (carbonyl cyanide m-chlorophenylhydrazone), and FCCP (carbonyl cyanide 4-(trifluoromethoxy)phenylhydrazone). CCCP and FCCP are lipid-soluble weak acids, which are much more powerful than DNP, as the mitochondrial uncoupling agents.

Both CCCP and FCCP are used at concentrations less than 1 μM. In my experiments, I usually used these uncouplers in the range of 0.3-0.5 μM. It should be remembered that because CCCP and FCCP are highly hydrophobic, the uncoupling concentration is based not on the incubation volume, but on the amount of mitochondrial protein. For the actively respiring heart, brain and kidney mitochondria the approximate working concentration of CCCP is 0.1 μM per 0.1 mg of protein.

CCCP (MM 204.6)   FCCP (MM 254.2)

However, if mitochondria are very active and the substrates are efficient, this approximate [CCCP] might not be sufficient to cause full uncoupling of oxidative phosphorylation. Therefore, the exact uncoupling concentration can be determined by titration with a given concentration of mitochondria, and using substrates, whose oxidation will not be inhibited upon addition of the uncoupler. For liver, brain and heart mitochondria this is the mixture of succinate 5 mM + glutamate 10 mM + malate 2 mM.

To mitochondria respiring in the State 4, you add an aliquot of the uncoupler, and after a minute, add the 2d aliquot. You will see what is happening. If after the 2d addition of the uncoupler there was a decrease or no change in oxygen consumption, than the first dose of the uncoupler was sufficient for full uncoupling. If the 2d addition of the uncoupler caused acceleration in respiration, than you have to use the higher concentration. Remember, that if you use palmitoyl-carnitine as a substrate (which is dissolved in ethanol), you have to use more CCCP to fully uncouple respiration because ethanol and Palmitoyl-carnitine will bind some amount of the uncoupler. This is registered as a drop in $\Delta\Psi$ while respiration may not increase.

Commentary 7. CCCP and FCCP are light sensitive, therefore the stock and working solutions have to be kept in bottles or tubes made of dark plastic or glass. I prefer CCCP over FCCP, because FCCP is less stable than CCCP. Prepare 0.65 mM stock solution of CCCP using high quality ethanol (see also Application 3 at the end of the book). Store the stock solution at -20ºC. For the current work, dilute the stock solution 10 fold (final concentration 0.065 mM), and take 0.5 ml sample into a 1.5 ml dark plastic tube, which you also store in the freezer. During the experiment, keep the working solution of CCCP on ice and tightly closed. Addition of 3-5 µl of this solution to 0.3 mg of brain mitochondria is close to optimal uncoupling concentration.

Commentary 8. Some reserchers add BSA to the incubation medium. BSA will also bind significant amount of uncouplers. Therefore you have to titrate the uncoupler in order to finde the effective concentration. The best solution would be not to use BSA at all.

Commentary 9. After addition of the aliquot of CCCP to the chamber, the oxygen trace may shift a little to the left. This is because the added cold hydrophobic solution contains oxygen.

Commentary 10. Recent review on the effects of uncouplers on mitochondria: Lou P-H., Hansen B.S., Olsen P.H., Tullin S., Murphy M.P., Brand M.D. (2007) Mitochondrial uncouplers with an extraordinary dynamic range. Biochem. J. 407(Pt. 1), 129–140.

As I have mention above, in Real Life, mitochondrial respiration far too often does not respond by increased oxygen consumption upon addition of the uncoupler. Here I discuss several most common situations: **1.** Addition of CCCP to mitochondria oxidizing succinate without rotenone causes inhibition of oxygen consumption in the State 3U. This is because upon deenergization of mitochondria, succinate dehydrogenase increases 10-fold its affinity to oxaloacetate, which is a powerful competitive inhibitor of the enzyme (Vinogradov et al. 1972). **2.** In brain mitochondria oxidizing glutamate + malate, addition of CCCP may cause inhibition of respiration because transport of glutamate into mitochondria, catalyzed by the glutamate-aspartate antiporter (aralar) is electrogenic. **3.** Deenergization of mitochondria also inhibits oxygen consumption supported by palmitoyl-carnitine. Transport of palmitoyl-carnitine and reactivation to acyl-CoA are also energy dependent events.

Thus, at the end of the polarographic experiment, you may encounter the situation when oxygen consumption becomes very slow, and it will take many minutes before the trace will reach zero $[O_2]$. In this case, it is practical to add 20-30 μl of Dithionite, which will quickly bring $[O_2]$ to zero. Be quick to wash both the chamber and the electrode with ample amount of water.

Commentary 11. Keep in mind that Dithionite is unstable. You keep the bottle with the dry powder in the dark, tightly closed. Prepare the solution of Dithionite every day – just add small amount (on a tip of a scalpel) into a 1.5 tube of dark plastic, add 0.5-0.7 ml of water and vortex. Keep tightly closed. The solution gradually becomes inactive, therefore after 2-3 hours prepare a fresh solution.

CHAPTER 10

Correct Analysis of the Polarographic Chart; Calculation of Respiratory Rates, Respiratory Control Ratios and some Other Parameters

Preparing the polarographic chart for calculations

It is easy to analyze and calculate the results using the graphic recorded by the paper chart recorder. On the paper chart you indicate major information regarding the current experiment: Date, type of incubation medium, protein concentration, type of mitochondria, and indicate all additions. Often, you have to return back to the chart to review the data. You cannot remember all the details of an experiment.

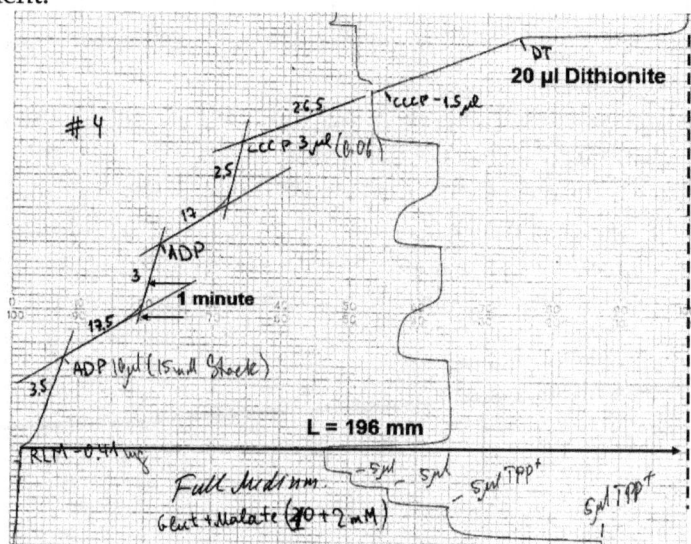

Figure 10.1. A typical polarographic chart recorded using a two channel paper chart recorder.

If you do not have the paper chart recorder, you must record your data using the computerized data acquisition software. The recorded experimental results are then used for creating a graphic similar to the one shown in Fig. 9.1. I found very inconvenient to create graphics in Origin.

During my carrier I used several different graphic softwares, such as Sigma plot, Slide Write and some older ones, which are now far behind the convenience of Origin. The major advantage of Origin is the possibility to present simultaneously on one chart several types of data, which have different Y scales. For the purpose of calculation, you indicate on the graphic the 60 second time grid (see Fig. 10.2).

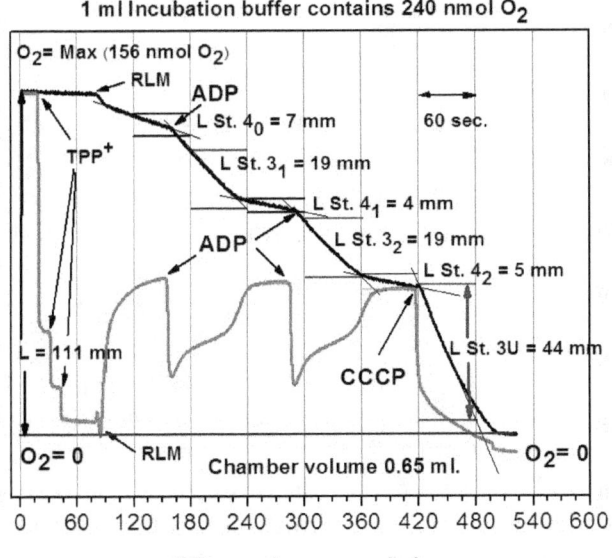

Time (seconds)

Figure 10.2. The graphic analysis of the polarographic trace obtained in experiment shown in Fig. 9.1. The measurements in mm were taken from the printed Origin graphic.

On the printed chart draw lines, as shown in Figures 10.1 and 10.2, paying attention that the lines cross the 1 minute grids.

Measure in mm the distance between the maximum and the zero concentrations of oxygen, which we designate as **L**, and the distances (L) between 1 minute cross points for each line representing the metabolic state. The calculated results from the above charts are presented in the Table 10.1 below.

Step 1. Calculation of the oxygen content ($[O2]$) in the incubation chamber.

The incubation medium is based on KCl, therefore 1 ml of the buffer contains 240 nmol O_2. The incubation chamber in the experiment shown in Fig. 10.2 has volume (with installed stopper) of 0.65 ml. Therefore, in the chamber the oxygen content is: $[O_2]_{Chamber}$ = 240 x 0.65 = 156 nmol.

Step 2. Calculation of the oxygen "price" for 1 mm of the working scale.

The working scale (**L**) is the distance in mm between the points on the graphic for maximum and zero oxygen contents. In our example L = 111 mm. Therefore, 1 mm = 156 nmol O_2/111 mm = 1.405 nmol O_2 per 1 mm.

Step 3. Calculation of oxygen consumption rates for each metabolic state.

In the Table 10.1, the second column shows the lengths of the 1 min distances in mm for each metabolic state measured from the chart in Fig. 10.2. Therefore the corresponding rates in nmol O_2 per 1 minute will be: V in mm x 1.405 = V nmol O_2 per 1 minute (see the 3d column in Table 10.1).

Step 4. Normalization of the respiratory rates for 1 mg of mitochondrial protein.

The results of calculation obtained during Step 3 divide by 0.5 - the content of mitochondria in the chamber in mg. The results are shown in the column 4 of the Table 10.1 as nmol O_2 per 1 minute per 1 mg of mitochondrial protein. If you multiply these results by 2, you will convert the results to nanogram atom O/min/mg. Nanomol is for O_2, nanogram atom is for O.

Step 5. Calculation of the respiratory control ratios.

Table 10.1. Calculations of Metabolic Rates and Respiratory Coefficients.

L = 111 mm		Max $[O_2]$ = 156 nmol	$[O_2]$ in 1 mm = 156/111 = 1.405 nmol O_2	
Metabolic state	1 min rate (L) In mm	V nmol/min	V nmol/min /mg protein	RCR
State 4_0	7	9.84	19.7	V_3/V_{40} = 2.7
State 3_1	19	26.7	53.5	V_3/V_{41} = 4.8
State 4_1	4	5.6	11.2	V_{3U}/V_{31} = 2.3
State 3_2	19	26.7	53.5	
State 4_2	5	7.1	14.2	
State 3U	44	62	124.0	

Historically, the naming of the mitochondrial metabolic states goes back to early 50s of the last century. The Table 10.2 shows designations of the metabolic states as defined by Chance & Williams (1955).

Chance and other researchers of those legendary years had their reasons to incubate mitochondria without substrates, wait until the endogenous substrates were exhausted in the presence of high ADP, and only then added a substrate, which initiated the metabolic State 3. After the added ADP became phosphorylated, the mitochondria were back in the metabolic State 4. For evaluation of the mechanisms controlling mitochondrial respiration, Chance and others offered a parameter, which we know as the respiratory control ratio, defined as the ratio State3/State 4 (Chance 1977). According to Chance, State 4 is the metabolic state after phosphorylation of the exogenous ADP, which in our experiments corresponds to State 4_1, State 4_2, etc. According to Lardi, the RCR is the Ratio of State 3/State 4 before addition of ADP.

Table 10.2. State of respiratory pigments in mitochondria. From Chance & Williams (1955) J. Biol, Chem. 217, 409-427.

Metabolic States →	State 1	State 2	State 3	State 4	State 5
Characteristics	Aerobic	Aerobic	Aerobic	Aerobic	Anaerobic
ADP level	Low	High	High	Low	High
Substrate Level	Low	Close to Zero	High	High	High
Respiration Rate	Slow	Slow	Fast	Slow	Zero
Rate-limiting component	Phosphate acceptor	Substrate	Respiratory chain	Phosphate acceptor	Oxygen

Of course, it would be not only impractical in our days to follow the Chance's experimental design, but erroneous, because we have other tasks – to evaluate the functional states of the mitochondria, excluding possible complicating situations, which were unknown 70 years ago. Take, for example State 1, now we know that brain and spinal cord mitochondria have no endogenous substrates. In State 2, the endogenous substrates approach zero, and we know that this will lead to conversion of the mitochondrial ATP and ADP to AMP, therefore the addition of a substrate, which will initiate oxidative phosphorylation (State 3), will not reach the maximum rate of oxidative phosphorylation.

I made this brief historical reminiscence for one thing, to pay tribute to the Classics of the Mitochondriology, whose terminology we still use, and for the other, I want the reader to understand, what is behind the names of the metabolic states of mitochondria. This will help us in discussions of the parameters, which are widely used, but not always correctly interpreted.

Different types of State 4

When we calculated the respiratory control ratios from the data presented in Fig. 10.2, we found that for the ratio of State 3_1/State 4_0 the RCR was 2.7, while for the ratio State 3_1/State 4_1, the RCR was 4.8. What happened? Why such a big difference in the ratios? Were mitochondria first uncoupled and then suddenly became coupled again? To answer these questions, we have to scrutinise each of the metabolic states, as did Chance & Williams in the Table 10.2 above.

State 4_0. First, we should keep in mind that these are liver mitochondria. If you work with mitochondria from other organs, you have to consider their peculiarities as well.

In the metabolic State 4_0, mitochondria are provided with oxygen, substrates, but there is no ADP, or other adenine nucleotides outside mitochondria. Mitochondrial inner membrane is impermeable for AN. Transport of ADP and ATP during oxidative phosphorylation occurs via specific ATP/ADP carrier - adenine nucleotide translocase (ANT), which works as a pore with two binding sites for ADP and ATP on the cytosolic and the matrix sides of the membrane. The binding sites have the same affinity for free ATP and ADP, but do not bind AN complexed with Mg^{2+}. The unidirectional transport of ADP from the cytosol to matrix and active pumping out of ATP during oxidative phosphorylation occurs for two reasons: 1) at normal pH 7.2 in the cell, ATP and ADP exist as polyanions, correspondingly as ATP^{4-} and ADP^{3-}. In energized mitochondria the outer surface of the inner membrane has positive charge, therefore the extra negative charge of ATP^{4-} is actively pumped out of the mitochondria, and thus helps to generate high ATP/ADP ratio in the cytosol. 2). Outside mitochondria, most of ATP^{4-} is complexed with Mg^{2+}, forming two isomeric forms of $MgATP^{2-}$ (Fig. 10.3).

Therefore, the concentration of free ATP^{4-} outside mitochondria is very low, and ADP^{3-} has preference in binding to ANT and being transported into the matrix in the energized mitochondria. Though, at high concentration of ADP and ATP (at about 0.5 mM), the adenylate kinase reaction may come into play, and the situation becomes more difficult to control and interpret.

ANT acquires one of the two conformational states only if it is bound to either ADP or ATP. If bound to ADP^{3-}, the ATP/ADP carrier changes conformation from the "c" (cytosolic) to "m"-conformation (matrix). In the matrix, ADP^{3-} becomes converted by the ATP-synthase to ATP, without being released into matrix, and ANT binds ATP^{4-}. The ANT-ATP^{4-} changes conformation and acquires the "c"-conformation, releasing ATP^{4-} into the cytosol. In the cytosol ATP^{4-} turns into $MgATP^{2-}$, which cannot bind to ANT.

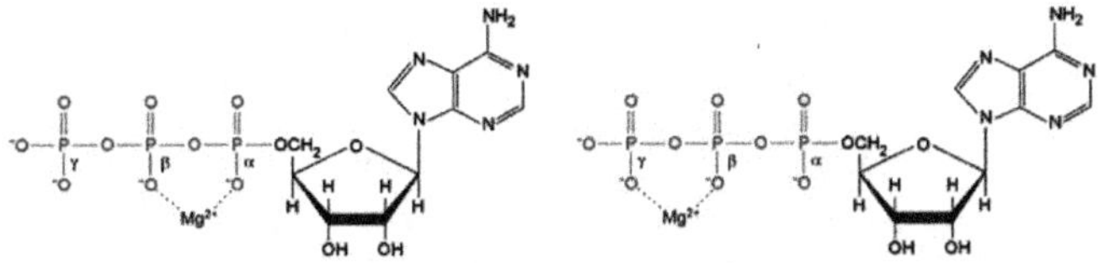

Figure 10.3. Two isomeric forms of the ATP-Mg2+ complex. ATP has alpha phosphyl group, beta phosphyl group, and gamma phosphyl group.

State 4_0. In State 4_0 there is no ADP outside mitochondria, therefore most ANT molecules "freeze" in the "c"-conformation, waiting for ADP to bind and change conformation to the "m" state. In the "c" conformation, the ANT works as a channel for K^+ and H^+ (Panov et al. 1980). Because the matrix space is more alkaline than the extramitochondrial space, the H^+ moves through the c-ANT channel into matrix and thus cause slight uncoupling. Because ANT is one of the most abundant proteins in the inner membrane, flow of H^+ into matrix discharges membrane potential, and the rate of the State 4_0 increases. That is why the ratio State 3_1/State 4_0 is significantly lower, than the State 3_1/State 4_1. With the brain and heart mitochondria this mechanism of State 4_0 uncoupling is usually not evident, as well as with slightly damaged liver mitochondria. This is more likely because of slight leakage of the mitochondrial ADP from mitochondria. However, because ANT has very high affinity to ADP, ADP binds to ANT, and changes conformation to "m" state.

We have also to keep in mind, that the liver mitochondria, in the experiment shown in Fig. 10.2, were isolated from the starved overnight animals. This means that the liver's metabolism was using fatty acids as substrates, and activated acyl-CoA have very high affinity for binding to ANT competing with ADP. Therefore RLM from the starved animals have most ANT in the "c"-conformation. In liver mitochondria isolated from the "Fed" animals, when the liver's metabolism is glycolytic, the differences in the rates of oxygen consumption between the State 4_0 and State 4_1 are negligible.

State 4_1. With the metabolic State 4_1 that is after phosphorylation of added ADP and at high ATP/ADP ratio outside mitochondria the situation is different. In State 4_1 outside mitochondria there is always some ADP present. Since ANT has very high affinity for ADP, the ANT is in the "m" conformation, and the conductance of the ANT channel for H^+ is low. In State 4_1 there may be several different situations that depend on the type and quality of the mitochondria. Above, we have discussed the situations for the liver mitochondria, when State 4_0 was higher than State 4_1. It seems, however, that with mitochondria from other organs this situation may not occur. On Fig. 6.1 you can see that with brain mitochondria oxidizing glutamate + pyruvate + malate, the States 4_0 and State 4_1 were almost the same. Similar situation can be observed also with the heart and kidney mitochondria. Once again, we see that liver mitochondria are cardinally different from mitochondria of other organs.

Because of the presence of ATP outside mitochondria, in State 4_1 the rate of respiration may be significantly higher as compared with the State 4_0, and the transition from State 3_1 to State 4_1 may be not sharp. This situation may occur if mitochondria from the skeletal muscle, heart or kidney were not purified in the Percoll gradient. The presence of damaged mitochondria stimulates the ATPase activity, which converts part of ATP to ADP, thus increasing the rate of the State 4_1. Addition of oligomycin - the inhibitor of mitochondrial ATPase, will inhibit this mechanism, and oxygen consumption in State 4_1 will slow down. However, it would be much better to purify mitochondria.

Mitochondria isolated from the hearts subjected to chronic hypoxy, may show very low rates of State 4_0 and State 4_1. This is the result of accumulation in the matrix of the heart mitochondria of long chain acyl-CoAs. In the heart, oxidation of fatty acids occurs via the carnitine–dependent pathway. After acyl-carnitine is transported to the matrix side of the heart mitochondria, acyl-CoA is formed, which enters into the β-oxidation pathway. In the hypoxic heart oxidation of fatty acids is low, and acyl-CoAs accumulate in the matrix. Long chain acyl-CoAs bind to ANT with very high affinity (K_i = 0.5 µM). From the matrix side, acyl-CoA fixes ANT in the "m" conformation and thus slows down the permeability of protons and thus help to maintain energization of mitochondria at low [O_2]. This is interesting adaptive mechanism.

The rates of respiration in metabolic state 4, regardless of the presence of AN, depend also on the reverse electron transport. Reverse electron transport (RET) stimulates ROS formation on Complex I, and is the energy consuming function. High rates of RET are observed with substrates and substrate mixtures that reduce the mitochondrial membrane pool of coenzyme Q. These are succinate, fatty acids and α-glycerophosphate. This situation is discussed in Chapter 12.

The ratio State 3U/State 3

Theoretically, both State 3 and State 3U should be close to each other because both metabolic states stimulate deenergization of mitochondria. But in the Real Life, the respiratory rates in State 3 and State 3U strongly depend on the type of mitochondria, type of substrate and the state of the experimental animal (metabolic state for the liver, presence of pathology).

Figure 6.1 shows, that in brain mitochondria oxidizing glutamate + pyruvate + malate the State 3U/State 3 ratio is close 1.0. This means, that in the brain mitochondria, the enzymes involved in oxidative phosphorylation are not rate limiting. For the liver mitochondria oxidizing glutamate + malate, the State 3U/State 3_1 was 2.3. This can be interpreted that in liver mitochondria the rate of the State 3 is limited by the oxidative phosphorylation enzymes. This conclusion agrees with the physiological functions of the liver mitochondria – unlike mitochondria from other organs, in liver mitochondria production of ATP for the cytosol is not the primary function. In the liver cells, many ATP-consuming key reactions of anabolic and catabolic pathways take place in the mitochondrial matrix (urea synthesis, synthesis of purines, etc.). On the other hand, we have already discussed that with some substrates the uncoupling of mitochondria resulted in inhibition of the State 3U. For example, for the rat heart mitochondria oxidizing palmitoyl-carnitine + malate, the State 3U/State 3 ratio = 0.3. I have found that the State 3U/State 3 ratio is a useful parameter for understanding of the mitochondrial metabolism in conjunction with the real rates of respiration only.

The above examples led us to conclusion that without analysis of the respiratory rates in different metabolic states, the respiratory control ratios are meaningless.

The ADP/O ratios

This ratio reflects the efficiency and the number of the "coupling sites" of the mitochondrial respiratory chain and ATP synthesis. Theoretically, with the NAD-dependent substrates there are 3 coupling sites, whereas with succinate as a substrate there are 2 coupling sites. The terminology reflects the ideology of the "classical chemical" hypothesis of oxidative phosphorylation. In the terminology of the Chemiosmotic theory, the "coupling sites" roughly correspond to "proton pumps".

Some researchers routinely use ADP/O ratio as indices of the "efficiency" of oxidative phosphorylation in mitochondria. Whatever the significance of this index, to my mind, it may not be worth the trouble, if you care about conducting the measurements of the ADP/O ratios correctly.

You cannot just add ADP and calculate how much O_2 was consumed between the moments of ADP addition and the transition from State 3 back to State 4. Some of the complication we have already discussed above. Here are some more "simply technical" complications:

First, you have to be sure that the isolated mitochondria are not "contaminated" with the damaged mitochondria. Damaged mitochondria have ATPase activity;

Secondly, commercial adenine nucleotides are far from being "pure". With the use of HPLC I have found that in some samples of ADP, the content was in fact a mixture of AMP (up to 40%), adenosine (up to 10%) and only the rest was ADP. In one case, instead of ADP, shown on the label, the bottle contained 100% AMP. I tested more than 5 different samples of ADP and ATP from different companies, and all of them were not even 90% close to the stated content. Therefore, before conducting the ADP/O measurement, you have to know exactly how much your sample of ADP contains ADP and AMP.

For the ADP/O measurement you have to use low concentrations of ADP, because at relatively high concentration of ADP, the mitochondrial adenylate kinase may "spoil" you measurements. To avoid this complication, you have to use P1,P5-di(adenosine 5') pentaphosphate, the inhibitor of adenylate kinase.

Therefore, in most experiments I did not use the ADP/O ratio for analysis of mitochondrial functions because the test is too technically demanding, with little contribution for understanding.

From the point of view of mitochondrial metabolism and physiology, the ADP/O ratio is meaningless, because in the Real Life mitochondria consume mixtures of substrates; some of them involve production of α-ketoglutarate, which produces ATP at the substrate phosphorylation step. Thus mitochondrial respiratory chain receives electrons from different point of entry, which are difficult to consider and evaluate for the ADP/O ratio.

Conclusions. From the above discussion, it is clear that polarographic assay, when conducted correctly, provides the most important information about the mitochondria. The analysis of the polarographic charts and the calculated rates and indices must not be mechanistic, but be always in conjunction with considerations regarding the origin of the mitochondria and the metabolic situations and the animal's metabolic phenotype.

CHAPTER 11

Measurements of $\Delta\Psi$ in Mitochondria with the TPP$^+$-sensitive Electrode

Preparation of the TPP+-sensitive electrode

First, you have to make the TPP$^+$-sensitive membrane. Table 11.1 shows the ingredients necessary for preparation of the TPP$^+$-sensitive membrane in the glass Petri dish of 48 mm diameter.

Table 11.1.

Ingredients	D=48 mm
PVC (Polyvinylchloride resin)	72 mg
Bis(2-ethyhexyl) Sebacate, 94% liquid (polymerizer) (BES). Sigma S3125; FW=426.7; d=0.92 g/ml.	131.4 µl
Tetraphenylboron (borate) NaBPh$_4$	1.25 mg
Tetrahydrofuran (THF)	2-3 ml

Commentary 1. Tetraphenylboron (TPB) NaBPh$_4$ sold by Sigma was of a very bad quality (at least during the period of 1990 -2010). The membranes made with Sigma's TPB would work just for few days. I used TPB from Aldrich, and the electrodes remained sensitive for months.

Commentary 2. CCl$_4$ and THF are very TOXIC! Work in the hood, and store the freshly-made membrane between two sheets of waxed paper placed in a book.

Dissolve the components in THF by using magnetic stirrer and do not shake. The little glass vessel in which you prepare the THF solution must be tightly closed. Open it only for addition of more THF. You may add more THF from the beginning (the final volume is not very essential), and after the components become completely dissolved (transparent), pour the solution into the Petri dish and leave covered with the lid in the hood to evaporate very slowly. After drying (usually next day), the membrane will be soft and easily detach from the Petri dish. First, using a sharp small scalpel detach the membrane around the perimeter, and then with the help of two small pincers carefully remove the membrane from the dish. Place the membrane between two sheets of thick waxed paper (as for weighing). Store under light pressure (inside a book) to keep it evenly flat. The membrane can be used for years.

Assembling the TPP$^+$ electrode

For the electrode you can use glass disposable microsampling 200 µl pipettes. After cutting the glass pipette of suitable length, dull the sharp end of the tube with an abrasive or over the flame of the gas burner. Otherwise the tight Tygon tube will be difficult to fit. Put on the glass capillary tube a piece of the Tygon tube (15-20 mm long). The working end of the tube, to which the membrane will be glued, must be even and smooth.

Remember, all procedures with Tetrahydrofuran (THF) MUST be performed under hood. Place a small (30-40 µl) drop of THF on the glass surface of the Petri dish cover and smear the edge of the tube with THF by rubbing it on the wet surface, then apply on the membrane sheet with the removed upper film of waxed paper. Apply light pressure and hold steady for 30-40 seconds. To cut the electrode with the glued TPB membrane, use a cork cutter tube slightly larger than the outer diameter of the electrode's tube (Fig. 11.1). Using a magnifying glass, check the integrity of the membrane glued to the electrode. If some parts of the membrane on the electrode are loose you can stick the membrane's edges to the tube by applying a tiny amount of THF with a micropipette. Be careful not to expose the inside of the PVC tube to THF for a long time, because the tube will swell and then crack. Therefore, during gluing the membrane use only a small drop of THF spread thinly over the glass surface. THF evaporates very rapidly; therefore the procedure of gluing of the membrane should be performed quickly.

Let the electrode dry overnight. Fill the capillary tube with 10 mM TPP$^+$ dissolved in 120 mM KCl (for better electrical conductance). Place another 10-15 mm piece of PVC tube on the top tip of the glass capillary tube with inserted connector (male) soldered to a thin platinum wire. You can use platinum wire from old discarded chambers for electrophoresis (Fig. 11.1). The soldering tin must have no trace of silver. Before use, soak the new electrode in 2-3 drops of 10 mM TPP$^+$ solution overnight and rinse with water before installing into the incubation chamber. To avoid an air bubble inside the capillary at the membrane, shake the electrode as the mercury thermometer in the old days.

Female part of the connector is soldered to the reference Ag/AgCl electrode. Place a regular commercial Ag/AgCl electrode into a 50 ml tube through a hole in the cover. Fill the tube with 15 ml of 3M KCl. To connect the reference electrode and the incubation chamber use a PVC tube of the same diameter as for the TPP$^+$-electrode tip. The cover of the 50 ml tube should have the second hole for the PVC tube. How to make the connecting KCl-Aga-agar bridge was described earlier.

Both the TPP$^+$ electrode and the connector can work for months. The TPP$^+$ electrode can work even for years. Between experiments store the TPP$^+$ electrode in a glass tube with a few drops of 10 mM TPP$^+$, and the bridge's glass tips filled with the KCl-agar gel - in the 3 M KCl solution. Both storage tubes have to be sealed with the parafilm to prevent drying. If you plan to store the TPP$^+$ electrode for a long time, store it dry, and disassemble the bridge tube and store the agar-KCl tips in the 2 M KCl solution (the agar will remain swelled).

To renew the electrode, fill the capillary tube with fresh 10 mM TPP$^+$ dissolved in 120 mM KCl and activate the electrode by leaving the electrode's tip (overnight) in the few drops of 10 mM TPP$^+$ solution.

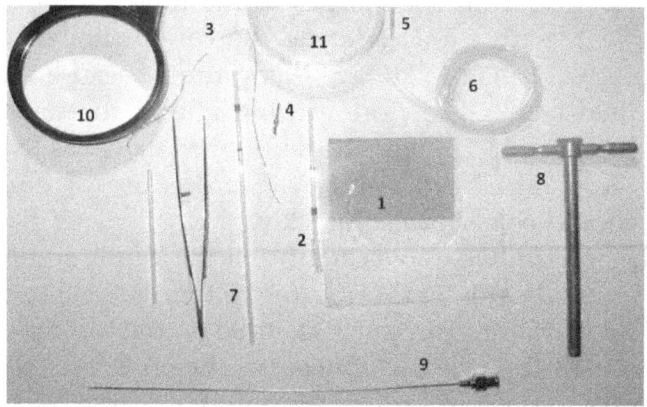

Figure 11.1. A. Items necessary for assembling the TPP$^+$ electrode. 1. The TPP$^+$-sensitive membrane; 2. Electrode's body; 3. Platinum wire; 4. Mail contact; 5. Female contact; 6. Tygon PVC tube; 7. 200 µl disposable pipette; 8. Cork cutter; 9. Long needle for filling the electrode with 10 mM TPP$^+$ solution; 10. Magnifying glass; 11. Glass Petri dish cover.

IMPORTANT. When working with the membrane-based electrodes, like the TPP+-sensitive electrode, pH electrode, and Ca^{2+}-sensitive electrodes, remember that when using in the same incubation chamber two electrodes, for example a TPP+ electrode and a membrane pH electrode, each electrode MUST have a separate Ag/AgCl reference electrode in their circuits and a separate bridge.

To minimize noise, wrap a piece of aluminum foil around the electrode assembled in the chamber, and connect to the ground by a low resistance copper wire using a crocodile connector. The pH meter and the magnetic stirrer also have to be grounded. For grounding you can use the nearest cold water pipe in the laboratory.

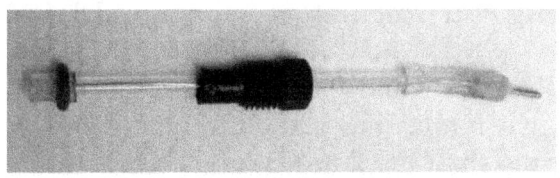

Figure 11.2 B. The fully assembled TPP+ electrode for polarographic chamber. On the tip with the TPP+-sensitive membrane there is a rubber ring and the screw, which fixes the electrode to the respiration chamber (see Fig. 8.12). The rubber ring is smeared lightly with the vacuum grease.

In more than 90% of the electrode's failure or excessive noise, the cause lies in bad connections. These may be an air bubble in the measuring electrode, or reference electrode, or connecting bridge tube, loose connections, and detached platinum from the connectors. The latter cause happens because tin and platinum to not mix and KCl is a very aggressive compound, which easily penetrates and damages the soldered parts. Even if the platinum looks to be tightly soldered to the connector, at the site of contact with platinum the surface of the metallic β-tin over time becomes converted to the gray non-metallic α-tin, which has no conductance for electrons. So, when you have noises, do the following: First check the contacts and connectors. Re-solder platinum wire to the new connector. Rather often, the BNC connectors fail unexpectedly. Therefore have several spare BNC connectors to connect your instruments with the computerized Data Acquisition system. Check the TPP+ electrode and connector bridge for air bubbles.

The PVC based membranes of the TPP$^+$-sensitive or Ca^{2+}-sensitive electrodes are hydrophobic. Therefore, additions to the incubation chamber of reactants, such as inhibitors, dissolved in ethanol, DMSO, or other hydrophobic solutes will affect the membranes, and washing of the electrodes with ethanol will strongly affect their performance. Therefore, if you must add to the incubation medium some components dissolved in ethanol or DMSO, add them in a minimal volume before the calibration with the TPP$^+$.

The TPP$^+$-sensitive electrode will response not only to tetraphenyl phosphonium, but also to the mitochondria targeted compounds, which contain the triphenylphosphonium moiety, such as Mito-TEMPO.

Measuring and Calculating Mitochondrial Membrane Potential

The TPP$^+$ electrode and the reference electrode are connected to the pH-meter, which must have an output for recorder. For recording mV changes using the Data Acquisition, the pH meter, set at the mV mode, must be connected first to the amplifier. The gain must be set for 100. The output of the amplifier is connected to the computer or the chart recorder.

According to Kamo et al. (1979):

$$\Delta\Psi = 2.303 \times RT/F \times \log(v/V) - 2.303\, RT/F \times \log[10F\Delta E/2.303RT-1]$$

where **v** is the mitochondrial volume, V - incubation medium volume, ΔE deflection of the TPP electrode from the base line prior to injection of mitochondria.

Table 11.2. Working solutions of TPP$^+$.

Tetraphenylphosph onium chloride	Mol. Wt. 374.85
10 mM TPP$^+$ solution	374.85 mg/100 ml 100 mM KCl
0.1 mM TPP+	100 µl of 10 mM TPP$^+$ per 10 ml water

When measuring membrane potential in mitochondria, first wash the electrode thoroughly with water and leave it for about 20 minutes in the relatively large volume (about 5 ml) of buffer under stirring. This procedure will remove excess of TPP$^+$ from the surface of the electrode after the storage in 10 mM TPP solution and increase sensitivity of the electrode. The electric potential is strong enough to be measured at the recorder's sensitivity set at 100-200 mV. At [TPP] =0 there may be some drift of the pen. If the drift is too large, this indicates that the electrode was not washed thoroughly. After equilibration of the initial trace, add three or four consecutive aliquots of 0.5 μM TPP, and wait for about 1 minute after each addition in order to obtain a steady-state line as shown in figure 11.3 below. The final concentration of TPP will depend on the amount of mitochondria and their respiratory activity. The responses of the electrode to additions of TPP follow the logarithmic scale and you should avoid adding too much TPP when the trace does not change or the response is too small upon addition of a new aliquot of TPP+. The upper practical limit is about 2.5-3 μM.

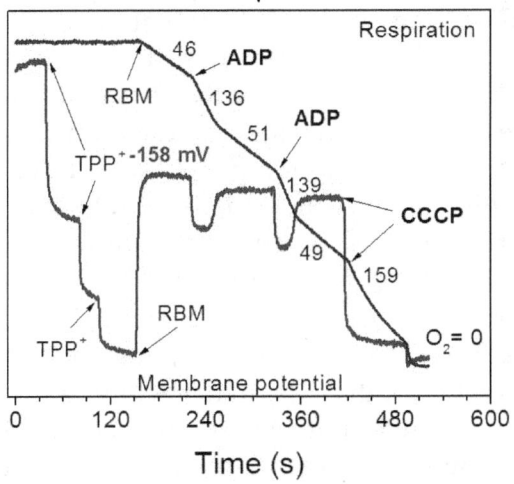

Time (s)

Figure 11.3. Calibration with TPP+ and changes of the TPP+ electrode signal during changes in respiratory rates. Incubation conditions: This figure is a replica of Figure 6.1. **Incubation conditions:** The incubation medium contained 10 mM glutamate + 2.5 mM pyruvate + 2 mM malate, final volume 0.65 ml. **Additions:** RBM 0.3 mg, ADP 150 μM, CCCP 0.4 μM. The numbers show the rates of O_2 consumption in nanomol O_2 per minute per 1 mg mitochondrial protein. The numbers at the membrane potential trace show the values of calculated $\Delta\Psi$ in mV as determined by the TPP+- sensitive electrode. TPP+ was added as 1.0 μM aliquot in the 1st addition and the next two as 0.5 μM aliquots, final concentration 2.0 μM. The relatively low $\Delta\Psi$ (-158 mV) was more likely associated with high reverse electron transport with this substrate mixture (The State 4_0 was high), which dissipated membrane potential and produced ROS at high speed.

145

Remember also, that TPP. TPMP+ or Mito-TEMPO accumulate in mitochondria several orders of magnitude over the concentration of these compounds outside mitochondria. To my experience, the optimal (outside, before addition of mitochondria) concentration is 1.5 -2.0 μM TPP. After addition of mitochondria the remaining TPP+ signal outside mitochondria should be at about no less than ½ of the signal after addition of the 1st aliquot of TPP+ (1.0 μM). It is better to diminish the amount of added mitochondria if too much TPP+ was consumed by the mitochondria. In the experiment shown in Fig. 11.3 the final concentration of TPP+ was 2 μM. The first addition was 1 μM TPP+, and the following two additions were 0.5 μM TPP each.

It is very important that maximal consumption of TPP+ would not reach or was too close to the zero point of the calibration. The State 4_0 steady-state trace should be at least at the half of the first trace after TPP+ addition (Fig. 11.3). The incubation media, in which the membrane potentials are measured, have to have a sufficiently high electrical conductance. Therefore, the KCl-based media are optimal. It is possible to measure $\Delta\psi$ in the sucrose medium without added KCl but in the presence of at least 2 mM potassium salts of Pi.

Before addition of mitochondria, calibrate each experiment by adding TPP in 0.5 μM aliquots. The first addition can be 0.5-1.0 μM, and then by adding 0.5 μM TPP up to final concentration of 1.5-2.0 μM.

For calculations of the membrane potential shown in Fig. 11.3 for the rat brain mitochondria, the matrix volume was taken as 0.5 μl for 1 mg RBM. For the liver mitochondria the accepted matrix volume is 1 μl for 1 mg. In my experiments with normal rat liver mitochondria oxidizing succinate (without rotenone) the calculated State 4_0 value for $\Delta\Psi$ varied between -185 and -165 mV.

According to Fontaine et al. (1998) the residual Donnane potential after full depolarization with FCCP was estimated to be -80 mV. In this $\Delta\Psi$ range the TPP+ or TPMP+-selective electrodes have too low sensitivity for a precise determination of $\Delta\Psi$.

Calculation of the Mitochondrial Membrane Potential

Russell Scaduto, with whom I worked at the Hershey Medical Center (Hershey, Pennstate University) in 1992-1994, developed computerized calculation of the mitochondrial membrane potential using the Excel spreadsheet with formulas, which make corrections for binding of TPP+ with the outer and inner compartments of mitochondria. The description of these calculations in the Excel's spreadsheet is presented in Table 11.3. For rat liver mitochondria (RLM) the External Binding Constant (EBC) = 55; the Internal Binding Constant (IBC) = 7.45; for the rat heart mitochondria (RHM) the EBC = 35, the IBC = 27.7 (LaNoue et al. 1986b).

ΔpH can be estimated with loading mitochondria with BCECF (2',7'-bis)carboxyethyl)-5(6)-carboxyfluorescein). The method is described in Bernardi et al. (1992). Because I never used this method, I cannot give any practical advice on measurements of ΔpH. However, you should keep in mind, that ΔΨ has very low capacity and therefore instantly responds to changes in energization of mitochondria. Mitochondrial ΔpH, on the other hand, is strongly buffered and thus essentially remains constant during regular manipulations with mitochondria under conditions that incubation medium contains anions, such as inorganic phosphate, acetate, etc.

Here are some references on measurements of ΔΨ and ΔpH: Rottenbberg 1984; Kamo et al. 1979; LaNoue et al. 1986; Jensen et al. 1986; Labajova et al. 2006.

Table 11.3. Calculation of the mitochondrial membrane potential using Excell template.

Column	Significance and formulas
A	Designates Experiments
B	Length in mm between zero [TPP$^+$] and at 0.5 μM [TPP$^+$]
C	Length in mm between zero [TPP$^+$] and at 1.5 μM [TPP$^+$]
D	Slope =(LOG10(1.5)-LOG10(0.5)/(C18-B18)
E	Interception = Log10(1.5)-D18*C18
F	Measured length in mm between zero [TPP$^+$] and Experimental point.
G	Amount of TPP outside Mt as log10 at the point: = F8*D8+E8
H	Medium (outside Mitochondria) = [TPP$^+$]ext in nmol/ml = 10^(G8)
I	Fraction of external TPP+ bound to mitochondria. In the Formula TPP+ itself was eliminated. The initial

		Formula is Fraction, here Frn = EBC*[TPP]ext*[Mt]/ ([TPP]ext+ (EBC*[TPP]ext*[Mt])). Here, we are taking into account concentration of mitochondria. This amount will be considered. In the cell's formula refer [Mt] to the cell in the column K when we will calculate [TPP+]in.
J		Amount of external TPP+ bound to Mitochondria (Mt) externally. Formula: H9*I9
K		Concentration of mitochondria ([Mt]) in mg
L		Matrix volume. Since for RHM and RLM it is assumed that 1 mg Mt has matrix volume of 1 mcl, than it numerically is equal to [Mt]
M		Amount of TPP+ in the mitochondria: =1.5 - H9 - I9. That is from the initial [TPP+] = 1.5 µM we subtract the amount of [TPP]ext, measued directly with the TPP-electrode, and amount of TPP, which is bound externally.
N		Mitochondrial [TPP+]in expressed in nmol/ml: (M9/L9)*1000
O		Fraction of [TPP+]in bound to Mt inside: (IBC*[Mt])/(1 + (IBC*[Mt])).
P		Matrix [TPP]free: N9 * (1-O9). Since total amount of TPP=1.0, and fraction bound is O9, then fraction of [TPP+]free is 1-O9.
Q		Definition of an Experiment. Here for convenience.
R		P9 is free [TPP+]in and H9 is free [TPP+]ext, measured directly with the TPP+ electrode.

As with all calculations using Excel spreadsheets, you must very carefully check the cells for correct values and formulas.

CHAPTER 12

Production of Reactive Oxygen Sspecies (ROS) in Mitochondria

Reactive Oxygen Species (ROS) are reactive molecules and free radicals derived from molecular oxygen. Correspondingly, Oxidative Stress is a generalized term for designating oxidative damages to proteins, lipids and DNA, and related functions caused by ROS.

General considerations regarding Oxidative Stress and the roles of Mitochondria in ROS production

In spite of intensive research on the mechanisms of ROS generation and prevention of oxidative stress, there are still huge gaps in our knowledge on these problems. On the one hand, this is because we still do not know many basic principles that govern ROS generation and control the defense against oxidative stress. To a large degree this is because we often do not take into consideration the tissue and cell-specific peculiarity of mitochondrial metabolism, function and location inside the tissue. On the other hand, many researchers are "blinded" by old paradigms regarding mitochondria, often use erroneous methods, such as unjustified usage of plate readers or wrong indicators for registering ROS production (Fridovich 1997). In this chapter I tried to compile information scattered in the literature about the basic principles of ROS generation and to discuss what we do know about metabolic and functional diversity of mitochondria in different tissues, and how this affect ROS generation and the methods of their detection.

In living organisms production of the oxygen-based radicals is inherently associated with the properties of the molecular oxygen and the mechanism of aerobic respiration of mitochondria. Molecular oxygen (O_2) has the electron structure that makes oxygen susceptible to radical formation, and O_2 is chemically very active. Oxygen is much better soluble in lipids than in water (Dzikovski et al. 2003; Desauniers et al. 1996; Möller et al. 2005; Subczynski and Hyde 1983). Therefore, biological membranes not only are no barrier for oxygen, but the concentration of O_2 is higher in the lipid phase of the membranes than in the cytosol.

When the dehydrogenases strip electrons from hydrogen, the electrons are carried one by one along the chain of oxidoreductase enzymes containing transition metal(s), such as Fe-S clusters, to the terminal enzyme of the respiratory chain – cytochromoxidase (Complex IV), where electrons, H^+ and O_2 interact to form H_2O in a highly exorgenic reaction. At some points along the respiratory chain, electrons may "jump" on O_2 with the formation of oxygen radical $O_2^{\bullet-}$.

It is useful to remember that although the $O_2^{\bullet-}$ radical is called "Superoxide radical", the term "super" does not imply exceptional reactivity. The term "super" was designated to the potassium salt of the radical anion KO_2 and had nothing to do with the chemical reactivity of $O_2^{\bullet-}$ (Sawyer and Valentine, 1981). Because mitochondria contain active superoxide dismutate (SOD2), in a cell most of the oxidative hazards are probably associated with hydrogen peroxide (H_2O_2) and the radicals derived thereof, such as lipid radicals or hydroxyl radical ($OH^{\bullet}$) (Murphy 2009). However, $O_2^{\bullet-}$ can directly damage proteins containing 4Fe-4S cluster (aconitase, Complex I, Complex II), or indirectly via formation of the highly toxic peroxynitrite anion ($ONOO^-$).

In mitochondria, $O_2^{\bullet-}$ is the first radical with one extra electron. Some enzymes, such as xanthine oxidase, can add the second electron to $O_2^{\bullet-}$ resulting in the formation of hydrogen peroxide (H_2O_2). Xanthine oxidase, and several other enzymes, can produce simultaneously both $O_2^{\bullet-}$ and H_2O_2, depending on reaction conditions (Kelley et al. 2010). Mitochondrial $O_2^{\bullet-}$ however is dismutated by SOD2 to hydrogen peroxide (H_2O_2), which can freely diffuse from mitochondrial matrix into cytoplasm.

Because ROS were forming in the living organisms since the development on Earth of the oxygen-containing atmosphere, oxidative stress was the problem for all organisms, particularly the aerobic ones, from the very beginning of their evolution. During billions years of evolution aerobic organisms developed various ways of protecting themselves from oxidative stress. These include various types of enzymes that eliminate ROS, inhibition of enzymes that stimulate production of ROS, and, finally, ROS themselves begun to serve as messenger signals to trigger defensive mechanisms against oxidative stress. Some cells, such as macrophages, even use superoxide radicals as a "weapon" to protect the host organisms from bacteria (Szatrowski and Nathan 1991).

Mechanisms of superoxide radical generation by mitochondria

Properties of molecular oxygen. The concentration of molecular oxygen ($[O_2]$) in air-saturated aqueous buffer at 37°C is around 240 µM (Reynafarje et al. 1985). Rough estimates for the mitochondrial $[O_2]$ *in vivo* are in the range 3–30 µM (Turrens 2003). However, these calculations of $[O_2]$ were made for the water milieu of the cell. In the cellular membranes the actual $[O_2]$ may be significantly higher, which also depends on a number of conditions that determine solubility and diffusion of O_2 (Möller et el. 2005; Windrem and Plachy 1083). The partition coefficient of O_2 between membranes and water (K_P) was found to be 4.9 ± 0.8 for dimyristoylphosphatidylcholine liposomes (Möller et el. 2005), and for the more complex erythrocyte plasma membrane the K_P = 4.73 (Power and Stegall 1970). Möller et al. (2005) have stressed that O_2, ·NO, and other apolar molecules, are less favorably dissolved in the polar region of the phospholipids in periphery of the membrane and should therefore be excluded (as well as proteins) from the hydrophobic volume. In comparison with liposomes, low density lipoproteins (LDL) dissolve lower amounts of oxygen due to a more ordered distribution and a higher packing of cholesterol esters and triglycerides in the core of LDL (Möller et el. 2005). When compared with organic solvents, lipid membranes and lipoproteins show additional constraints to O_2 dissolution related to the acyl-chain ordering that suggest a lower value for K_P than in an equivalent isotropic phase (Möller et el. 2005). In this respect it is important that in comparison with the cytoplasmic membrane, which has the highest level of cholesterol, the mitochondrial membranes have only a trace of cholesterol and the highest content of phosphatidylethanolamine (Kopeikina-Tsiboukidou and Deliconstantinos 1983; Madden 1980). This is possibly related to the lower affinity of cholesterol for phosphatidylethanolamine. It was shown that cholesterol has a preferential affinity for neutral phospholipids in the following order: sphingomyelin > phosphatidylcholine > phosphatidylethanolamine (Demel et al. 1977). Cholesterol has been shown to decrease the fluidity of membrane phospholipids in the liquid crystalline state, and this higher packing of lipids excludes O_2 from the membrane (Möller et el. 2005).

Thus, the lipid phase of the mitochondrial inner membrane should have the highest content of dissolved O_2 in comparison with other membranous structures in the cell. Changes in the fatty acid composition and their saturation affect solubility of O_2 in the membranes, and in some species serve as an adaptive mechanism to the changes in ambient temperature (Desaulniers et al. 1996).

In addition to the partitioning effect, the diffusional properties of the reactants also influence the reaction rates in the lipid milieu. It can be observed that O_2 diffusion is nearly twice that of $\cdot NO$ (another biologically relevant radical), and both molecules diffuse in the core of LDL ≈ 10 times slower than in water. A consequence of this is that the diffusion of $\cdot NO$ and O_2 in lipoproteins and membranes will be less sensitive to changes in viscosity because of varying cholesterol content or fatty acid unsaturation degree. On the other hand, the increase in lipid packing induced by cholesterol can lead to a lower $\cdot NO$ or O_2 solubility in a lipid milieu (Möller et el. 2005). The product of partition (K_P) and diffusion (D) coefficients, $Do\alpha = K_P \times D$, is a good indicator of the rate at which diffusion-controlled reactions will occur in a lipid milieu (Windrem and Plachy 1980). The results indicate that $Do\alpha$ is greatest in the bilayer center and least near the bilayer head groups. The properties of the inner mitochondrial membrane suggest that both solubility and diffusion of O_2 are highest as compared with other cellular membranes.

Formation and properties of superoxide radical ($O_2^{\cdot -}$)

The figure below lists the major species of oxygen-containing radicals.

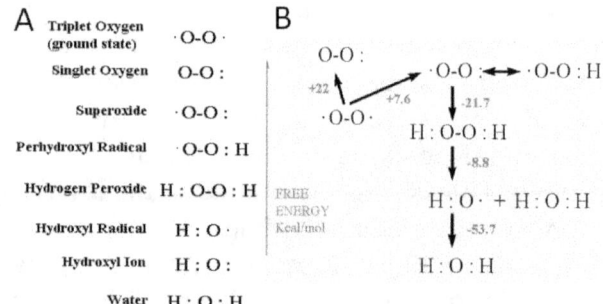

Figure 12.1. Nomenclature of various forms of oxygen (A) and the energy required for activation states of oxygen (B).

Note that although formation of $O_2^{\cdot -}$ is endothermic, the subsequent transformations to H_2O_2, hydroxyl radical (HO$^{\cdot}$) and H_2O are irreversible and highly exorgenic. Energy of the reactions are in

Kcal/mole. (From "Oxidative stress" by Bryan D. McKersie, University of Guelph; posted on Internet in 1996).

Properties of the superoxide anion radical

The superoxide anion radical ($O_2^{•-}$) is the first to originate in mitochondria during normal respiratory activity, and is commonly considered as the origin of all other species of oxygen-containing radicals, with the exception of nitric oxide and semiquinone radical. Within mitochondria, $O_2^{•-}$ is produced by the one-electron reduction of O_2, which is controlled by the kinetic and thermodynamic factors determining the interaction of potential one-electron donors with O_2 (Murphy 2009). The presence of two unpaired electrons in antibonding orbitals (that is the most energetic and remote from the nucleus orbitals) with parallel spins makes the ground-state O_2 to accept one electron at a time (Sawyer and Valentine 1981). The standard reduction potential for the transfer of an electron to O_2 to form $O_2^{•-}$ is −160 mV at pH 7 (Sawyer and Valentine 1981), for a standard state of 1 M O_2. As the pKa of $O_2^{•-}$ is 4.7, this standard reduction potential is invariant across most biological pH values (Murphy 2009; Sawyer and Valentine 1981). The actual reduction potential (Eh) that determines the thermodynamic tendency of O_2 to form $O_2^{•-}$ will vary with the relative concentrations of O_2 and $O_2^{•-}$ (Murphy 2009; Sawyer and Valentine 1981).

Eh (mV) = −160 + 61.5 log10[O2]/[O2•−] (Reaction 1)

The steady-state intramitochondrial $[O_2^{•-}]$ is difficult to measure accurately, but it is likely to be very low because of different from O_2 physical-chemical properties of $O_2^{•-}$ (Sawyer and Valentine 1981), and the presence of micromolar concentrations of MnSOD (SOD2) in the mitochondrial matrix (Murphy 2009).

Superoxide is a relatively small univalent anion with an O-O bond distance that is intermediate between that for O_2 and H_2O_2. Its solvation energy in water indicates that it will form exceptionally strong hydrogen bonds with H_2O (Sawyer and Valentine 1981). The effect of the strong solvation of $O_2^{•-}$ by water result in exclusion of the $O_2^{•-}$ from the lipid phase of the membrane, and the superoxide radical, in the form of hydroxyperoxyl radical, will spontaneously dismutate in the reaction: $2 O_2^{•-} + 2H^+ \leftrightarrow O_2 + H_2O_2$.

with equilibrium in aqueous media far to the right ($K_{pH7} = 4 \times 10^{20}$) and thus dismutation of $O_2^{•-}$ by H_2O is complete. However, in the mitochondrial matrix there is enzyme superoxide dismutase - MnSOD

(SOD2), which will very rapidly catalyze dismutation of $O_2^{\bullet-}$ to H_2O_2 by a process that is first-order with respect to $[O_2^{\bullet-}]$ and is several orders faster than the spontaneous dismutation (Murphy 2009).

Thus, *in vivo*, the one-electron reduction of O_2 to $O_2^{\bullet-}$ is thermodynamically favored, and both physical properties of $O_2^{\bullet-}$ and the presence of MnSOD will also favor the reaction of O_2 with electron donors to form $O_2^{\bullet-}$, in effect increasing the flux of electrons from electron donors to H_2O_2 (Andreev et al. 2005; Murphy 2009). However, only a small proportion of mitochondrial electron carriers with the thermodynamic potential to reduce O_2 to $O_2^{\bullet-}$ generate superoxide radical. Under most circumstances, small-molecule electron carriers such as NADH, NADPH, $CoQH_2$ (reduced coenzyme Q) and glutathione (GSH) do not react with O_2 to generate $O_2^{\bullet-}$. According to Murphy (2009), mitochondrial $O_2^{\bullet-}$ production takes place at redox-active prosthetic groups within proteins, or when electron carriers such as $CoQH_2$ are bound to proteins. There are structural and kinetic factors that favor or prevent the one-electron reduction of O_2 to $O_2^{\bullet-}$ by mitochondrial electron carriers.

Hydroperoxyl radical. Molecular formula: HO_2; Molar mass: 33.01 g mol^{-1}.

The hydroperoxyl radical, also known as the perhydroxyl radical, is the protonated form of superoxide with the chemical formula HO_2. Hydroperoxyl is formed through the transfer of a proton to an oxygen atom.

Reactivity. The superoxide anion, $O_2^{\bullet-}$, and the hydroperoxyl radical are in equilibrium in aqueous solution: $O_2^{\bullet-} + H_2O$ is in equilibrium with $HO_2 + OH^-$. The protonation/deprotonation equilibrium exhibits a pKa of 4.88; consequently, about 0.3% of any superoxide present in the cytosol of a typical cell is in the protonated form. As was mentioned above, HO_2 is much less hydrophobic and thus excluded from the lipid phase of membranes. Unlike O_2^-, which predominantly acts as a reductant, HO_2 can act as an oxidant in a number of biologically important reactions, such as the abstraction of hydrogen atoms from tocopherol and polyunsaturated fatty acids in the lipid bilayer. As such, it may be an important initiator of lipid peroxidation.

Effect on environment. Because dielectric constant has a strong effect on pKa, and the dielectric constant of air is quite low, superoxide photochemically produced in the atmosphere is almost exclusively present as HO_2. As HO_2 is quite reactive, it acts as a "cleanser" of the atmosphere by degrading certain organic pollutants. As such, the chemistry of HO_2 is of considerable geochemical importance. Hydroperoxyl is responsible for the destruction of ozone in the stratosphere, and it is formed as a result of the oxidation of hydrocarbons in the troposphere.

The sites of superoxide radical formation in mitochondria

One of the most controversial and difficult problem regarding oxidative stress is determination of the sites and the mechanisms of $O_2{}^{\bullet-}$ production in mitochondria. This issue was discussed in a large number of papers and reviews, and here we reference on few of them (Murphy 2009; Barja 1999; Brand 2010; Kudin et al. 2008; St-Pierre et al. 2002; Turrens 2003).

The intramolecular conditions for $O_2{}^{\bullet-}$ generation. The mitochondrial electron transport complexes are multiprotein entities comprising many polypeptides carrying various cofactors and transition metals necessary to transfer electrons. According to Murphy, the most important factor that determines $O_2{}^{\bullet-}$ production by mitochondria is the proportion, P_R, of a given electron carrier that is reactive with O_2 to form $O_2{}^{\bullet-}$, and P_R responds rapidly to a range of biological situations (Murphy 2009). At the molecular level, the final factor affecting the rate of $O_2{}^{\bullet-}$ production by electron carriers within proteins is the second-order rate constant (kE) of their reaction with O_2, which depends on the distance between O_2 and the electron donor (Marcus and Sutin 1985). This is similar to electron movement down the respiratory chain, which occurs by electron tunnelling from carrier to carrier, with a maximum distance of approximately 14 Å (1 Å=0.1 nm) between each carrier for effective tunnelling to occur (Moser et al. 2006). Similar distance constraints apply probably to the reaction of protein-bound electron carriers in the respiratory chain with O_2 to form $O_2{}^{\bullet-}$, with the bulk of the protein acting as an insulator to keep O_2 at a safe distance from the carriers and thereby minimizing $O_2{}^{\bullet-}$ production (Moser et al. 2006; Murphy 2009; Ohnishi et al. 2005). Consequently, $O_2{}^{\bullet-}$ production will probably occur only at the sites where O_2 can approach closely to electron carriers, namely at active sites exposed to the aqueous phase, such as FMN of Complex I and FAD of Complex II, or to the membrane core were the protein-bound CoQ may interact with the membrane pool of CoQ (Murphy 2009). The membrane pool of CoQ serves as a cumulative sink for electrons coming from dehydrogenases oxidizing succinate, fatty acids or α-glycerophosphate, and thus may increase the redox state of the protein-bound CoQ. The partially reduced form of CoQ, such as a semiquinone, may be the critical electron donor (Murphy 2009).

In addition to the discussed above structural conditions, that determine at the intramolecular level production of $O_2^{\bullet-}$ inside the respiratory chain complexes, there are structural determinants that affect $O_2^{\bullet-}$ production at the macrostructural organization of the respiratory chain. The superstructural organization of respiratory chain determines not only the capacity of a given site to generate $O_2^{\bullet-}$, but also is responsible for the differences in the rates of ROS generation in mitochondria from different organs.

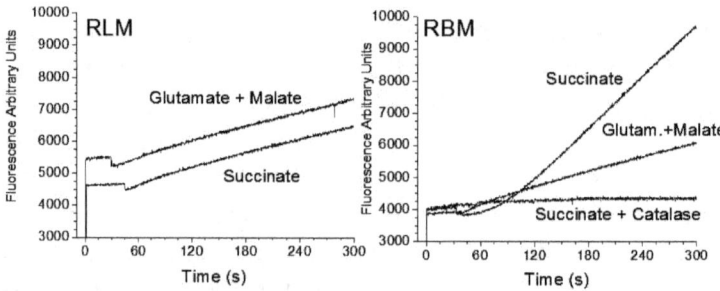

Figure 12.2. ROS generation by rat liver and brain mitochondria with glutamate + malate and succinate as substrates. ROS production was registered as H_2O_2 by Amplex Red method. Additions: RLM 0.5 mg/ml, RBM 0.2 mg/ml, glutamate 10 mM, malate 2 mM, succinate 5 mM, malonate 5 mM, Amplex red 2 μM, Horse radish peroxidase (HRP) 1 Unit, volume 1 ml. The Figure depicts both differences in the rates of ROS production (0.5 mg/ml RLM and 0.2 mg/ml RBM) and complications of the method, which will be explained in details in the Methods. Here, it is sufficient to say that the presence of catalase in RLM makes the Amplex Red method unsuitable for these mitochondria.

The concentration of the enzyme responsible for $O_2^{\bullet-}$ production, will vary with organism, tissue, state, age or hormonal status, and may underlie many of the changes in the maximum ROS production capacity between tissues (Barja 1999); for example, complex I content may explain the different maximum capacities of pigeon and rat heart mitochondria (St-Pierre et al. 2002), or between liver and brain (Panov et al. 2007) (Fig. 12.2).

The supercomplexes of mitochondrial respiratory chain and ROS generation

In mitochondria oxidizing the NAD-dependent substrates, the rate of ROS generation does not depend on the energy state of mitochondria, and the rate of ROS production is relatively slow (Panov et al. 2007). The latter is because with flavin containing dehydrogenases the rate-limiting step appears to be the initial 1-electron transfer (Massey 1994). Only when the electron flow was prevented by respiratory inhibitor rotenone, did the generation of ROS accelerate several fold due to increased reduction of the CoQ sites of Complex I (Votyakova and Reynolds 1991; Starkov and Fiskum 2003; Panov et al. 2007).

The steady low level of ROS generation occurs at the initial rate-limiting step which is FMN and thus strongly depends on the mitochondrial NADH/NAD$^+$ ratio (Murphy 2009; Kussmaul and Hirst 2006). Other potential sites of ROS generation in the respiratory chain, after the rate-limiting FMN of Complex I, are oxidized and thus do not produce $O_2^{\bullet-}$ (Panov et al. 2007). This is ensured by the superstructure of the respiratory chain.

It was shown that for the heart mitochondria the ratio for oxidative phosphorylation (OXPHOS) complexes I : II : III : IV : V is 1 : 2 : 3 : 6 –7 : 3–5 (Hatefi 1985), or more recently determined as 1 : 1.5 : 3 : 6 : 3 (Ohnishi et al. 2005). The respiratory complexes interact with each other to form a supercomplex named the respirasome (Schägger 2001). On the basis of the above ratios of the OXPHOS complexes, Schägger (2001) suggested that the respirasome exists as a mixture of two large supercomplexes and one smaller complex. Each of the two large supercomplexes are comprised of Complex I monomer, Complex III dimer, and four copies (two dimers) of Complex IV. The smaller supercomplex contains two of Complex III and four of Complex IV (Schägger 2001; Schägger and Pfeiffer 2000).

The major advantages of the supercomplex structure of the mitochondrial respiratory chain are substrate channeling, catalytic enhancement, sequestration of reactive intermediates and structural stabilization (Schägger 2001; Schägger and Pfeiffer 2000). The advantage of the substrate channeling is the use of localized substrate molecules, for example, quinone (CoQ) and cytochrome c, which can react independently of bulk properties of a quinone or cytochrome c pool (Schägger 2001).

We suggest that in addition to the benefits listed above, the respirasome is also an evolutionary adaptive mechanism designed to prevent excessive production of ROS. Evidently, this mechanism developed early during the evolution of the aerobic organisms because it is available in aerobic bacteria and yeast (Schägger 2001;

Schägger and Pfeiffer 2000). The initial reaction of NADH with the FMN of the Complex I is the rate limiting step in oxidation of the NAD-dependent substrates. The suggested composition of the respirasome ensures that all components proximal to Complex IV, which is in a 4-fold excess of Complex I, are kept oxidized regardless of the energy state of the mitochondria. In brain, kidney and skeletal muscle mitochondria, which respire at rates similar or even higher than heart mitochondria, the composition of the respirasome should be similar to that reported for the beef heart mitochondria (Schägger 2001; Schägger and Pfeiffer 2000).

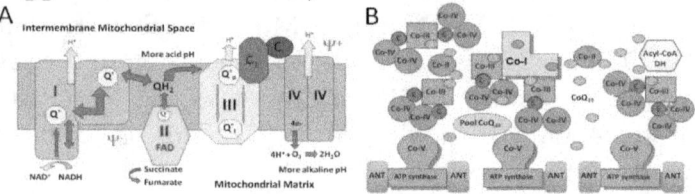

Figure 12.3. Schematic presentation of the mitochondrial respiratory chain. A. Traditional presentation of the respiratory chain. The scheme is true for a pathway of a single electron, but it misses the complex superstructure of the respiratory chain, which greatly affects the kinetics and the redox states of electron carriers. **B.** The supercomplex structure of the respiratory chain complexes – respirasome. This is a very simplified scheme because it does not take into consideration interactions of the respiratory enzymes with other functional structures of the tricarboxylic cycle enzymes, substrate transporters, aminotransferases and other mitochondrial enzymes.

There is another important aspect of the supercomplex structure of mitochondrial enzymes, which is often missed when considering mechanisms of ROS generation. The matrix of mitochondria from the heart, brain, skeletal muscle is a hard gel, which excludes or greatly hampers the diffusion of small molecules. Therefore, many enzymes of the tricarboxylic acid (TCA) cycle are tightly associated with the inner mitochondrial membrane. For example, the membrane-bound pools of pyruvate dehydrogenase complex (PDHC) and α-ketoglutarate dehydrogenase complex (KGDHC) about 200 times exceed the "soluble" pool of these enzymes in the matrix (Stanley et al. 1980). The enzymes are organized into stable functional complexes, which tunnel metabolites along the complexes forming efficient metabolic pathways. Such multienzyme complexes have been shown for β-oxidation of fatty acids (Angdisen et al. 2005), malate-aspartate shuttle and TCA cycle enzymes (LaNoue and Williamson 1971; LaNoue et al. 1972). In addition, mitochondria *in vivo* oxidize several different substrates simultaneously coming from different metabolic pathways, which are strongly organ specific (Panov et al. 2009; Panov et al. 2010b).

Determination of mitochondrial ROS generation sites by inhibitory analysis

A useful instrument to study the mechanisms of ROS generation and a contribution of a respiratory chain complex is the application of highly specific inhibitors of electron transport in the mitochondrial respiratory chain. Figure 12.4 shows schematically the structure of the respiratory complexes I, II and III and the sites of action of the electron transport inhibitors.

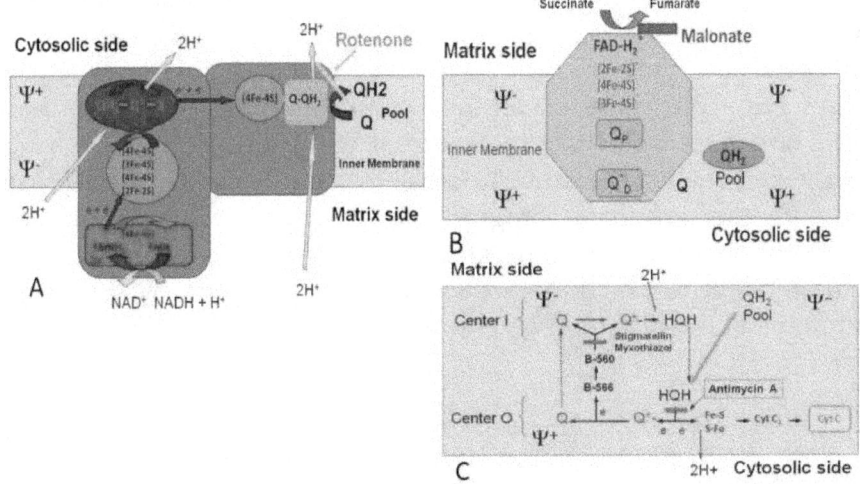

Figure 12.4. Schematic presentation of respiratory Complex I, Complex II (Succinate dehydrogenase), Complex III and the sites of action of the specific electron transport inhibitors. A. Complex I - Specific inhibitor rotenone (green bar); B. Complex II (SDH) - specific inhibitor malonate; C. Complex III – specific inhibitors stigmatellin and myxothiazol, which prevent reduction of the QI and QO, and antimycin A, which prevents transport of e- to cytochrome C1, and thus causes full reduction of the sites QI and QO.

Примером успешного применения ингибиторного анализа могут служить работы из лаборатории Мартина Бранда (Brand M. D.) по изучению продукции АРК митохондриями скелетных мышц (Quinlan et al., 2012, 2013; Perevoshchikova et al., 2013).

Generation of superoxide radicals during oxidation by mitochondria of the NAD-linked substrates. In the current literature, the authors traditionally consider pyruvate and glutamate as "classical" substrates for Complex I. However, in the brain and spinal cord mitochondria these substrates can be metabolized via the transaminase reactions with formation of considerable amount of succinate. In brain mitochondria from Sprague Dawley rats up to 50% of pyruvate were oxidized via the formation of succinate (Panov et al. 2009, 2012). Therefore, in the brain and spinal cord mitochondria a considerable amount of ROS during oxidation of pyruvate or glutamate is associated with the reverse electron transport (Panov et al. 2009, 2012).

These data suggest that in the course of investigation of ROS production by the isolated mitochondria it is necessary to take into account the metabolic features of the organ or tissue. Our data suggest also that significant phenotypic (genetic) differences exist not only between organs and different species, but also within the same species but from different strains. While working with the animal model of Amyotrophic Lateral Sclerosis, we have observed, at the end of our experiments, that mitochondria isolated from brains and spinal cords of the original Sprague Dawley rats begun to manifest significant changes in substrates metabolism. As a result, the hybrid animals failed to respond to the mutated *sod1* gene by developing symptoms of the disease (Panov et al. 2009, 2012).

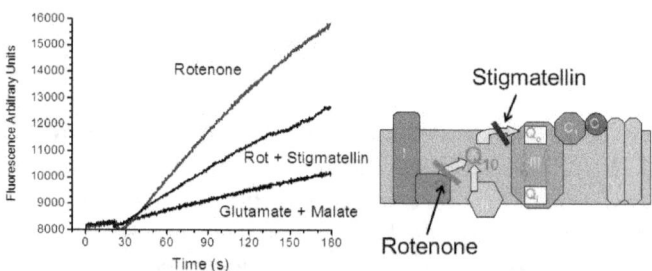

Figure 12.5. ROS generation by rat brain mitochondria oxidizing glutamate + malate. ROS production was registered as in Fig. 1. Inhibitors were added before mitochondria 0.2 mg/ml: rotenone 5 µM, stigmatellin 5 µM. Explanation in the text.

Figure 12.5 illustrates the well-known fact that addition of rotenone to the brain mitochondria oxidizing glutamate + malate results in a dramatic increase of ROS production. Rotenone prevents transport of electrons from Complex I down the respiratory chain, and reduces the sites that may generate superoxide radicals. Therefore, the researchers made a straightforward, now accepted as

"Classical", conclusion that Complex I is solely responsible for the increased production of ROS in the presence of rotenone.

Nevertheless, if electrons cannot overcome the rotenone block, the metabolites can do this. Therefore, the mechanisms of ROS generation in the presence of rotenone may be more complex. For example, if mitochondria were incubated in the presence of rotenone + stigmatellin, the rate of ROS production diminished two fold (Fig. 12.4). Stigmatellin acts after the site of action of rotenone, and hampers the entrance of electrons into the Q-cycle of Complex III. Therefore a two-fold drop in ROS production in the presence of rotenone + stigmatellin suggested that almost half of the superoxide radicals were associated with Complex II. We have shown earlier that Complex II in brain mitochondria is capable of producing superoxide radicals with succinate as a substrate in the presence of rotenone + stigmatellin (Panov et al. 2005).

In 2005, we have shown that brain mitochondria from Lewis rats produced ROS on FAD of Complex II with the rate of about 30 pmol/min/mg protein, which was similar to production of ROS on the FMN of Complex I in the presence of pyruvate (Panov et al. 2005). The experiment presented on Fig. 12.5 was performed with brain mitochondria from Sprague Dawley rats (Panov et al. 2009). Since the rate of ROS generation in the presence of rotenone + stigmatellin was significantly higher than with glutamate + malate alone, we can presume that in the presence of rotenone part of the glutamate metabolism was overdirected to transamination with production of first α-ketoglutarate and then succinate. The above facts reflect, probably, one of the mechanisms of adaptation of mitochondrial metabolism for overcoming the sudden inhibition of electron transport on Complex I. In some metabolic phenotypes in the absence of inhibitors mitochondria, during oxidation of "classical" complex I substrates, are capable of producing ROS on FMN of Complex I, FAD of Complex II, and probably, under certain conditions (hypoxia), on Complex III.

Quinlan et al. (2012) have shown that Complex II in mitochondria from skeletal muscles can produce ROS during both direct and reverse transport of electrons, when the electrons first reduce the membrane pool of CoQ and then Complex I. Production of ROS during reverse electron transport can be inhibited either at the level of FAD of Complex II, which is SDH, by malonate, or at the level of interaction of SDH with the membrane pool of ubiquinone (CoQ) by atpenin A5. However, atpenin A5 did not influence the direct pathway of succinate oxidation (Quinlan et al., 2012). This proved that production of superoxide radicals on Complex II occurs only on FAD.

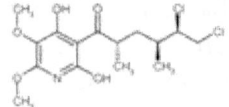 Atpenin A5 is a highly specific inhibitor of Complex II at the level of interaction of the enzyme with ubiquinone. Molecular mass 366.2. This is an antifungal antibiotic, which was initially isolated from the FO-125 Penicillium strain. A synthetic analog is currently available on the market. It is rather expensive – 1 mg cost more than $300.00. As is follows from the chemical structure, Atpenin A5 is a highly hydrophobic compound, which is soluble in acetone, chlorophorm, ethyl acetate, DMSO, methanol and 100% ethanol. Therefore Atpenin A5 must be added to the incubation chamber after mitochondria. However, the acting concentration is calculated based on the volume of the chamber. Currently, Atpenine A5 is the most powerful known inhibitor of Complex II with Ki = 3.6 nM!

Production of ROS during the reverse electron transport (RET). Until the recent publications, which have shown that Complex II can be one of the sites for production of superoxide radicals (Panov et al. 2005; Quinlan et al., 2013; Perevoshchikova et al., 2013), Complex II itself have never been considered as a source of ROS production. It was discussed only in terms of the succinate-induced ROS generation associated with the reverse electron transport.

Succinate dehydrogenase, being part of the tricarboxylic acid (TCA) cycle, was considered also as part of the respiratory chain named Complex II. However, other mitochondrial enzymes, such as acyl-CoA dehydrogenase, and sn-glycerol-3-phosphate dehydrogenase also supply electrons to the respiratory chain via the membrane pool of Coenzyme Q (Angdisen et al. 2005; Tretter et al. 2007c). All these enzymes, including SDH, reduce the membrane ubiquinone in irreversible reactions. These are the points of irreversibility, which make the corresponding metabolic pathways work only in one direction: TCA cycle works clockwise, β-oxidation of fatty acids produce acetyl-CoA, and the glycerophosphate shunt to pump protons from cytoplasm to the mitochondria. Strictly speaking, these enzymes are not part of the respiratory chain, but rather injectors of electrons into the respiratory chain using ubiquinone as an intermediate.

It is well-known that oxidation of succinate by mitochondria in the absence of rotenone increases several times production of ROS

(Votyakova & Reynolds, 2001; Panov et al., 2007; Starkov, 2008). This, however, is not the case if isolated brain and heart mitochondria preserved endogenous inhibition of SDH by oxaloacetate (that is brain mitochondria were isolated without BSA). To this day, some authors consider production of ROS during oxidation of succinate as a purely *in vitro* experimental phenomenon, which does not take place *in vivo* (Starkov, 2008). I have already wrote about these delusions in a number of my works (Panov et al., 2007, 2014), and we will discuss it more closely in the next section.

The tissue specific mitochondrial metabolism and ROS generation

ROS production by neuronal mitochondria. The significance of succinate in ROS production was not accepted by many researchers on the pretext that succinate concentration is too low in mitochondria (Starkov, 2008; Stowe & Camara, 2009; Zoccarato et al., 2007; Murphy 2009). However, we have recently shown that brain and spinal cord mitochondria produce succinate in the presence of pyruvate + malate (Panov et al., 2009), or glutamate + malate. Therefore, a large part (up to 50-60%) of ROS production by brain and spinal cord mitochondria with these substrates may be associated with oxidation of succinate formed within the multienzyme complexes (Panov et al. 2012). A particularly large production of the succinate-dependent ROS was observed in the presence of glutamate + pyruvate + malate (Panov et al., 2009, 2011a, 2011b). With these substrates the increased succinate oxidation was caused by a dramatic increase in mitochondrial α-ketoglutarate due to high activities of aminotransferase (Balazs 1965a, 1965b), and the release of SDH inhibition by endogenous oxaloacetate (Panov et al. 2009). It is should be kept in mind that the steady-state concentrations of metabolites may be irrelevant if they are formed inside the metabolic functional complexes, such as the complex formed by the glutamate transporter, glutamate transaminase and the TCA cycle enzymes (LaNoue and Williamson 1971; LaNoue et al. 1972). In these complexes the metabolites are formed and consumed without being released, particularly if the metabolic pathway is strongly irreversible due to coupling with SDH and the major point of irreversibility – Complex IV. Moreover, in the neuronal mitochondria succinate can be formed during catabolism of neuromediator γ-aminobutyric acid (GABA) (Tillakaratne et al. 1995). Thus, in the excited neurons succinate is an indispensable mitochondrial metabolite. Furthermore, we have shown that the succinate-dependent reverse electron transport is a subject for large phenotypic variations (Panov et al. 2012).

Another objection to the importance of the succinate-supported ROS generation argues that RET is energy-dependent and in the functioning mitochondria the diminished mitochondrial energization will inhibit production of ROS (Starkov, 2008). The latter objection is to a large degree valid for the most perpetually functioning organs, such as heart, kidney and liver. But it is only partially applicable to brain and spinal cord where most mitochondria are located at synaptic junctions (Wong-Riley 1989; Abeles 1991). In the narrow spaces of synaptic junctions there is no other task for the mitochondria besides provision of ATP for the restoration of ionic composition in excited synapses. If the neurons are not excited, mitochondria at synaptic junctions remain fully energized and may produce ROS. Therefore axonal and synaptic junctions, including neuromuscular junctions, are particularly vulnerable to oxidative damage if neurons are not excited for some reason.

We have found that formation of succinate during oxidation of glutamate, pyruvate, and, in particular, glutamate + pyruvate was specific only for the brain and spinal cord mitochondria. Spinal cord mitochondria in general oxidized succinate at higher rates due to the lower intrinsic inhibition of SDH, and they were more vulnerable to oxidative stress than brain mitochondria (Panov et al., 2011a, 2011b).

Reverse electron transport can be also activated during β-oxidation of fatty acids. Although it is known that neuronal mitochondria do not oxidize fatty acids, nevertheless, up to 20% of the overall energy expenditure in brain is covered by oxidation of fatty acids (Ebert et al. 2003), and brain tissue specifically accumulates carnitine and acyl carnitines (Nalecz et al. 2004). Although β-oxidation of fatty acids occurs in astroglial mitochondria providing energy for the anaplerotic synthesis of glutamate and glutamine, as well as saving glucose for production of lactate, we have studied the effects of palmitoyl-carnitine on mitochondrial respiration and ROS generation.

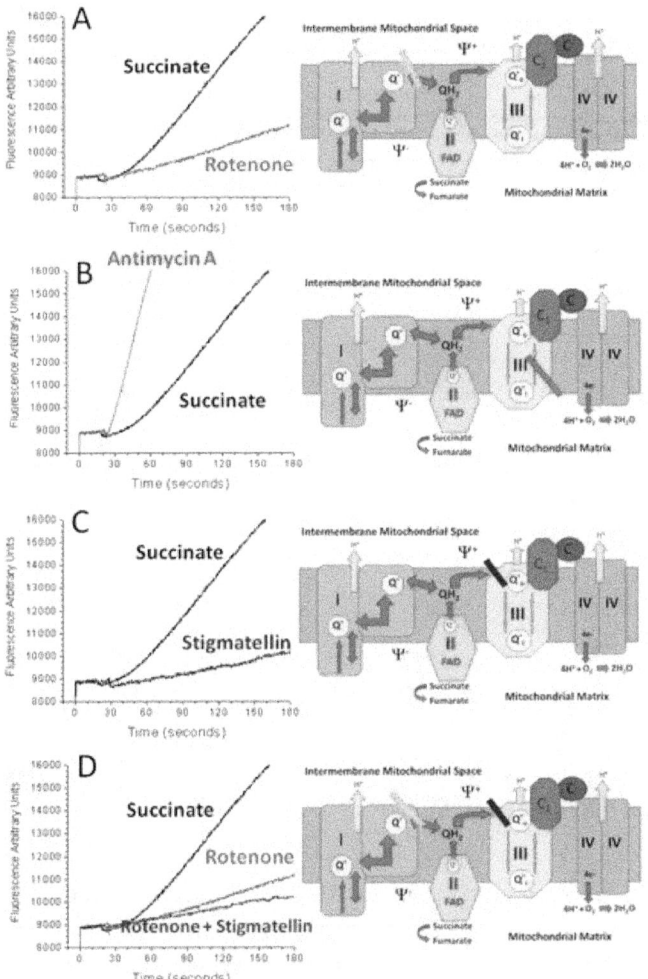

Figure 12.6. An example of Inhibitor analysis of the succinate-driven ROS generation by rat brain mitochondria. The analysis of situations during inhibitor analysis shows: 1) that during oxidation of succinate alone, most of ROS are produced on complex I (panel A); 2) Panel b indicates that in the presence of antimycin a large amount of ROS is formed on complex III, which is blocked by stigmatellin (panel C); 3) a comparison of panels C and D shows that in the absence antimycin but in the presence of rotenone only a small fraction of ROS is produced on complex III (the difference between rotenone and rotenone + stigmatellin); 4) In the presence of rotenone + stigmatellin ROS production occurs on complex II. Thus generation of ROS in the respiratory chain occurs at different sites and rates depending on the current situation.

In the experiments shown in Figure 12.7 rat brain mitochondria were isolated in the absence of BSA; therefore mitochondria preserved the intrinsic inhibition of SDH by oxaloacetate. Fig. 12.7 demonstrates that with succinate alone or palmitoyl-carnitine alone brain mitochondria had no stimulation of respiration upon addition of ADP or uncoupler (CCCP). However, when palmitoyl-carnitine was mixed with glutamate, pyruvate or succinate the mitochondria manifested high rates of oxidative phosphorylation and uncoupled respiration (Fig. 12.7).

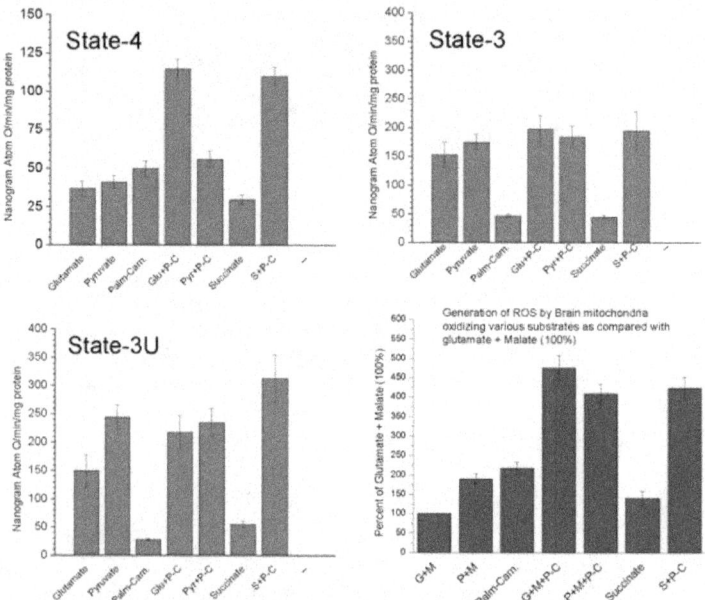

Figure 12.7. Respiratory rates and ROS generation in brain mitochondria oxidizing palmitoyl-carnitine, glutamate, pyruvate, succinate and their mixtures. Experimental conditions as described in Panov et al. 2007, 2009, 2011. Additions: glutamate 5 mM, pyruvate 2.5 mM, malate 2 mM, succinate 5 mM, palmitoyl-carnitine 25 µM, CCCP 0.3 µM; brain mitochondria 0.3 mg/ml.

The State-4 respiration with palmitoyl-carnitine mixed with glutamate or succinate increased several fold. In general, the rates of State 4 respiration correlated with ROS production. Figure 12.7 demonstrates that in comparison with glutamate + malate, the rates of ROS production with palmitoyl-carnitine mixed with glutamate, pyruvate or succinate increased 4-5 folds. These results give a new insight into the possible metabolic origin of neurodegenerative disorders observed in some people involved in high level of physical activities (Beghi et al. 2010) and metabolic syndrome.

ROS generation in heart mitochondria. The experiments presented above for the neuronal mitochondria and our experiments with the heart mitochondria suggest that a possible reason why the succinate-dependent ROS production was underrated could be the fact that most researchers use "classical" mixtures of substrates, such as glutamate + malate, pyruvate + malate, succinate alone, and very often, in the presence of rotenone. <u>In the brain and heart mitochondria SDH is normally inhibited by endogenous oxaloacetate</u> (OAA).

This inhibition is subject to phenotypic regulation and can be released by isolation of mitochondria in the presence of BSA (Panov et a. 2010b). However, in heart mitochondria the inhibition of SDH by OAA was insensitive to BSA, but can be released metabolically by addition of glutamate and/or pyruvate which remove OAA in transaminase reactions (Panov et al. 2009). It is conceivable that the physiological significance of this intrinsic inhibition of SDH is to prevent excessive ROS production associated with the reverse electron transport.

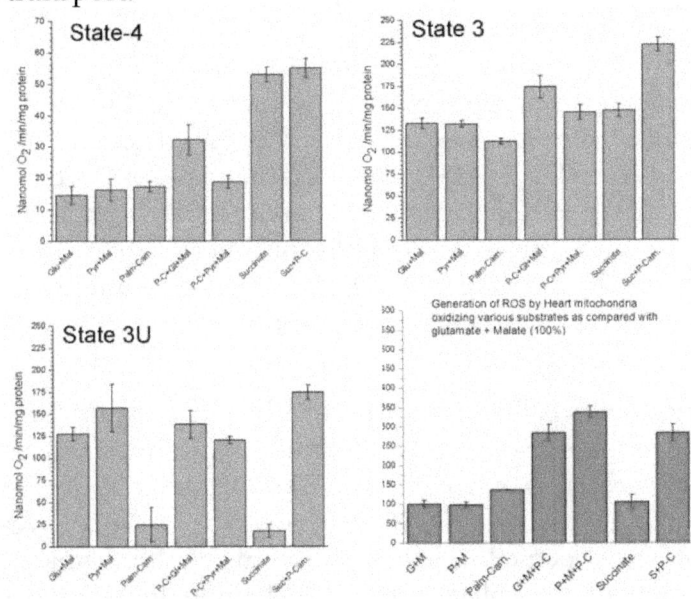

Figure 12.8. Respiratory rates and ROS generation in rat heart mitochondria oxidizing palmitoyl-carnitine, glutamate, pyruvate, succinate and their mixtures. Incubation conditions as in Fig. 12.7.

Figure 12.8 shows that simultaneous oxidation of palmitoyl-carnitine and glutamate, pyruvate or succinate led to several-fold increase in ROS production and significant increase in the rate of oxidative phosphorylation. The results presented for the substrate dependence of ROS generation by neuronal and heart mitochondria led us to conclude that the old methodology to study mitochondrial respiration and ROS generation using "classical" substrates does not reflect physiological conditions for mitochondria *in vivo*. To study ROS production *in vivo* is a very difficult and probably impossible task. Therefore when using mitochondria as a model to study mechanisms of oxidative stress, we have provide mitochondria with conditions that correspond to the organ's metabolic situation.

In this chapter we have provided evidence that argues that both *in vitro* and *in vivo* the major source of ROS is associated with the energy-dependent reverse electron transport (RET). The reduction of the membrane pool of CoQ, which drives the RET, may be associated not only with the activity of SDH (Votyakova and Reynolds 1991; Starkov and Fiskum 2003; Panov et al. 2007), but also with β-oxidation of fatty acids (Panov et al., 2011, 2014), and in some tissues α-glycerophosphate (Murphy 2009).

Indirectly the leading role of RET-dependent ROS generation in various pathologies was supported by numerous literature data on the ability of a wide range of agents and conditions, that inhibit RET, to delay or diminish oxidative stress (Brand et al. 2004; Dlaskova et al. 2008; Lambert et al. 2008; Miwa and Brand 2003; Skulachev 1996). Inasmuch as RET strongly depends on energization and intrinsic inhibition of SDH, the methods of isolation and the quality of isolated mitochondria are of paramount importance in studies of ROS generation by mitochondria.

Methods to study ROS production by mitochondria

Some researchers use cell cultures to study mechanisms, or the effects of antioxidants, such as MitoTEMPOL, on ROS production by mitochondria (Trnka et al. 2009). However, in many cases the researchers use penicillin and streptomycin to avoid bacterial contamination of the cell culture. It is known that aminoglycoside antibiotics (streptomycin, gentamicin) are mitotoxic (Avadhani and Buetow 1972; Davey et al. 1970; Wirmer and Westhof 2006). We have established that mitochondria isolated from a number of different cell lines grown in the presence of streptomycin do not respire on any substrate. Although mitochondria in cells grown in cultures containing antibiotics do respond to changing cellular conditions, they do not maintain aerobic metabolism, and glycolysis is the only source of ATP. Therefore many conclusions obtained on cell cultures with antibiotics have to be regarded with caution.

As discussed in Chapter 1, we have found that in order to isolate high quality mitochondria the chemicals used for preparation of the isolation and incubation buffers have to be of molecular biology quality and bidistilled water is the best for preparation of buffers. The Milli-Q water (a trademark created by Millipore) and water from similar systems, which use ion exchange resins, usually contain high or very high levels of hydroperoxides. We observed that some "high quality" systems produced water containing more than 100 nmol/ml H_2O_2 as determined with the Amplex red + horse radish peroxidase method.

High quality mitochondria have to show high rates of respiration as well as high respiratory control ratios (RCR = $V_{State 3}/V_{State 4}$). However, with some substrate mixtures, such as shown in Figures 12.7 and 12.8, the RCR values may be significantly diminished because of the high rates of RET. We have concluded that the rate of State 4 respiration is controlled not only by intrinsic proton conductivity, but also by the RET. Thus a decrease in RCR does not necessarily indicate on uncoupling and deterioration of the quality of mitochondria.

Measurements of ROS production

Direct measurements of $O_2^{\cdot-}$ inside mitochondria make little sense because of the presence in mitochondria of high activity of SOD2, which dismutate $O_2^{\cdot-}$ to H_2O_2 (Murphy 2009). Therefore the most reliable method to evaluate the activity of ROS generation in mitochondria is to follow the appearance of H_2O_2 outside mitochondria. Of the current available methods to determine H_2O_2, the Amplex red method is the most convenient and reliable, although the method also has some limitations and pitfalls.

Fluorimetric Measurements of Mitochondrial H_2O_2 by Amplex Red.

Storage and handling of chemicals and possible pitfalls. The Amplex Red (AR) is somewhat air sensitive, and is strongly light sensitive. AR must be stored at -20°C protected from light, desiccated. The AR is unstable in the presence of thiols, such as dithiothreitol

(DTT). If you plan to use DTT, the concentration of DTT must be not higher than 10 µM. The AR is also unstable at pH > 8.5. The absorption and fluorescence of resorufin is pH dependent. The measurements must be at pH range 7-8. The optimum is pH 7.4. At extremely high concentration of H_2O_2 (e.g. 10 µM, final concentration) AR can produce lower fluorescence than at moderately high levels, because excess of H_2O_2 can oxidize the reaction product, resorufin, to nonfluorescent resazurin.

Table 12.1. Concentrations of chemicals used for measurement and analysis of the ROS Generation Site.

Chemical or Inhibitor	[Stock] mM	Aliquot µl/ ml	[Final] µM
Amplex red (Invitrogen, MW 257.25)	0.4	5.0	2.0
Horse radish peroxidase	0.5 mg/ml		2.0-3.0 U/ml
Atpenin A5	0.0005	2.0	1.0
Myxothiazol	2.05	2.0	4.1
Rotenone	0.42	5.0	2.1
Stigmatellin	0.4	5.0	2
Antimycin A	0.8	2.0	1.6
CCCP	0.13	4.0	0.52
Resorufin (for calibration)	0.005	5.0	25.0 pmol/ml

For the sake of better controlling of your results, we recommend not to use AR and horse radish peroxidase (HRP) from the commercial kit "Amplex Red Hydrogen Peroxide/Peroxidase Activity Assay Kit" (A-22188) from Molecular Probes. To our experience the kits often are heavily "contaminated" with hydrogen peroxide and this creates problems. In addition, it is much cheaper to use commercially available Amplex red and HRP. I have avoded kits as much as possible. Kits are too expensive, and usually you have no control over what you are doing. There aresituations when you must use kits, but in most cases kits are for lasy and not very clever people.

For calibration, it is better use solutions of resorufin in ethanol (working stock concentration 5-10 µM, depending on the instrument's sensitivity). It is advisable to do calibrations at the end of every experimental day. Solutions of resorufin, under condition of storing in the darkness and refrigerated, are very stable. Calibrations with H_2O_2 basically give results similar to resorufin, but H_2O_2 is very unstable and preparation of usable stock solutions on everyday basis is very complicated, and in the process you may contaminate everything around with hydrogen peroxide.

Ethanol quality is very important. For preparation of resorufin and solutions of inhibitors and uncouplers use ethanol, which will give no fluorescence rise upon addition to the cuvette with buffer containing Amplex red + HRP. Very often, ethanol is dirty with fluorescent contaminants and peroxides. Resorufin for calibration is also light sensitive and should be stored in refrigerator and in the dark vessel. For preparation of the stock solutions of Amplex red (0.2 mM) and resorufin (10 μM) it is better to use ethanol. The stock solutions have to be diluted to a concentration in order to reach the desired final concentration in a minimal volume of addition. When you add to the incubation medium an inhibitor, uncoupler or other chemical dissolved in ethanol, keep in mind that by adding 5 μl of 70% ethanol to 1 ml of medium, the final concentration of ethanol (MW 46.07; Density 0.7893) will be 60 mM! Therefore, try to use ethanol at lowest concentration possible, that is minimize concentration as a solute (use 40 or 50% EtOH to prepare final working solutions), and the volume of addition. Similar considerations relate to other solutes, DMSO, for example. It should be kept in mind that ethanol and DMSO are scavengers of hydroxyl radical (Liochev 1996).

As we have mentioned above, serious problems arise from contamination of water and solutions with H_2O_2. Peroxides are extracted to water and solutions from the ion exchange resins and plastic containers. Therefore all solutions and ethanol must be stored only in glass containers. Filter fluorometers directly "see" contamination of solutions with H_2O_2 when using Amplex red-HRP method. With spectrofluorimeters you have to zero the instrument using water treated with Catalase. This will allow you to evaluate contaminations of your solutions with H_2O_2.

For fluorescence measurements, it is better to use disposable plastic cuvettes. Quartz cuvettes are expensive, and very difficult to wash from chemicals, which is a time consuming procedure. Do not prepare in advance samples of buffer containing substrates. Some substrates, such as pyruvate interfere non-enzymatically with H_2O_2. Therefore, after 15-20 minutes the background fluorescence will be different. In addition, air in big cities and buildings is often contaminated with Ozone and therefore give substantial rise to fluorescence even in the absence of mitochondria comparable to the rates observed with brain mitochondria oxidizing glutamate + malate. Therefore application of plate readers, which have large surface to volume ratio, may introduce errors.

Measurements of hydrogen peroxide with Amplex red method. The method is based on the reaction catalyzed by horse radish peroxidase: **Amplex Red + H_2O_2 → Resorufin**, which has strong fluorescence. The formation of resorufin is stoichiometric 1:1 with H_2O_2. For measurements of resorufin fluorescence use **Excitation wavelength 563 nm** and **Emission - 587 nm**.

Incubation conditions and measurement: the final concentration of Amplex Red 2 μM, plus 1-3 Unit/ml of horse radish peroxidase (Sigma), mitochondrial protein in the range of 0.05- 0.2 mg/ml. The reaction starts by addition of mitochondria. All other additions – substrates and inhibitors are added before mitochondria.

Controls: SOD (Sigma) 50 - 100 Units/ml, to check whether endogenous SOD2 has enough activity to convert all $O_2^{\bullet-}$ radicals to H_2O_2; Catalase (Roche Co. Do not use the enzyme from Sigma) 0.1 mg/ml, to destroy H_2O_2, and have Zero background for H_2O_2 content.

The Amplex red + HRP method works perfectly well with the heart, brain, spinal cord and kidney mitochondria. The method cannot be used with the isolated liver mitochondria because they contain high activity of catalase (Panov et al. 2011).

Commentary 1. In practice, weigh on analytical scales as little as you take from the bottle with HRP. This is a fluffy powder. For example, you picked up 0.3 mg. The volume of the buffer you have to add to obtain 0.5 mg/ml will be found from the proportion: 0.3 mg/X μl = 0.5 mg/ 1 ml; thus X = 0.3/0.5 = 600 μl. You can aliquot this volume into two tubes containing 0.3 ml each, and freeze one of them. The dissolved in the incubation buffer enzyme preserves its activity for several weeks when refrigerated. Working solution can be used for 2-3 days without freezing.

Commentary 2. In most spectrophotometers and spectrofluorimeters the beam of light goes too close to the surface of a buffer, if you use 1 ml. Therefore you have to use at least 2 ml of the buffer. However, you can safely use 1 ml of the buffer, if you place under the cuvette a plastic pad about 6-8 mm thick.

Commentary 3. Do not add ketoacids (pyruvate, oxaloacetate or α-ketoglutarate in advance because they interfere with H_2O_2 nonenzymatically in a general reaction: R-C=O COOH + H_2O_2 → R-COOH + CO_2 + H_2O, or for pyruvate

Pyruvate Hydrogen peroxide Acetic acid

$$\text{(structure)} + \text{(structure)} \rightarrow CH_3\text{-COOH} + CO_2 + H_2O$$

For this reason, the solutions of ketoacids as substrates are prepared freshly each working day.

Commentary 4. Since the most active productions of ROS are usually associated with the reverse electron transport, do not use EDTA during isolation of mitochondria or in the incubation media. As I have discussed above, EDTA removes Mg^{2+} from the high affinity sites in mitochondria, which triggers proton conductivity through the inner membrane. As a result, membrane potential drops by 30-40 mV and ROS production becomes inhibited. These adverse effects of EDTA can be reversed by the presence of 2-5 mM $MgCl_2$.

Other, than superoxide, radical reactive oxygen species of major biological significance

We present only a brief account on the major radicals, which have biological significance.

Hydroxyl radical. In living organisms there are two major reactive oxygen species, superoxide radical and hydroxyl radical that are being continuously formed and thus are potentially most harmful.

Hydroxyl radical Hydroxyl ion

· OH   OH⁻

The hydroxyl radical, OH·, is the neutral form of the hydroxide ion (OH-). Hydroxyl radicals are highly reactive and consequently short-lived. Hydroxyl radicals are produced from the decomposition of hydroperoxides.

(ROOH), in the atmospheric chemistry is formed by the reaction of excited atomic oxygen with water. It is also an important radical formed in radiation chemistry, since it leads to the formation of hydrogen peroxide and oxygen, which can enhance corrosion in coolant systems subjected to radioactive environments. Hydroxyl radicals are also produced during UV-light dissociation of H_2O_2 and in Fenton chemistry, where trace amounts of reduced transition metals catalyze peroxide-mediated oxidations of organic compounds. The hydroxyl OH• radical is one of the main chemical species controlling the oxidizing capacity of the global greenhouse gases and pollutants in the Earth atmosphere. It is the most widespread oxidizer in the troposphere, the lowest part of the atmosphere. Understanding OH• variability is important to evaluating human impacts on the atmosphere and climate. This oxidizing reactive species has a major impact on the concentrations and distribution of organic pollutants. The OH• species has a lifetime in the Earth atmosphere of less than one second. Besides the Earth's atmosphere, hydroxyl radicals are present in cosmic quantities in the interstellar dust (Wikipedia).

Biological significance. Among ROS, the hydroxyl radical is the most toxic and potentially most dangerous. The hydroxyl radical has a very short *in vivo* half-life of approximately 10^{-9} second and a high reactivity. Because the hydroxyl radical interacts with any molecule on the first encounter, it can damage virtually all types of macromolecules: carbohydrates, nucleic acids (mutations), lipids (lipid peroxidation) and amino acids (protein dysfunctions). At low level of hydroxyl radical formation it is of little danger because most of the encounters are with the non-essential molecules. However, during exposure to high amounts of ionizing radiation, or high level of H_2O_2 and the presence of divalent transition metals, formation of hydroxyl radicals causes structural damages to cellular components and DNA. Most biologically active hydroxyl radicals are being predominantly formed during the period of ischemia in which oxygen is in short supply (Lipinski, 2011).

Hydroxyl radicals can reduce disulfide bonds in proteins, specifically fibrinogen, resulting in their unfolding and scrambled refolding into abnormal spatial configurations. Consequences of this reaction are observed in many diseases such as atherosclerosis, cancer and neurological disorders. From experimental point, buffers prepared with Milli-Q water always contain high levels of H_2O_2, and sometimes (with old or bad cartridges) very low levels of Cu^{2+}, Fe^{2+} or Zn^{2+}, which will initiate Fenton reaction and thus hydroxyl radicals will deteriorate isolated mitochondria very rapidly. Cu^{2+} and Zn^{2+} have several orders higher activity in generation of hydroxyl radical in Fenton reaction than Fe^{2+} ions, and are active at concentration as low as 10^{-12} M (Brewer 2007).

Unlike superoxide, which can be detoxified by superoxide dismutase, the hydroxyl radical cannot be eliminated by an enzymatic reaction. Mechanisms for scavenging peroxyl radicals for the protection of cellular structures include endogenous antioxidants, such as melatonin and glutathione, and dietary antioxidants, such as mannitol and vitamin E (reviewed in Lipinski 2011). For experimental work, it is useful to remember that when you add a chemical to the incubation medium dissolved in DMSO, say 5 μl/ml, the final concentration of DMSO will eliminate hydroxyl radical and thus prevent the damages you may expect from your treatment.

Generation and Properties of Hydroxyl Radicals. In the Haber-Weiss reaction hydroxyl radicals are generated in the presence of hydrogen peroxide and iron ions. The first step involves reduction of ferric into ferrous ion: $Fe^{3+} + O_2^{\cdot -} \rightarrow Fe^{2+} + O_2$. (Halliwell, 1996). The second step is the Fenton reaction: $Fe^{2+} + H_2O_2 \rightarrow Fe^{3+} + OH^- + OH^{\cdot}$. The iron-mediated Haber-Weiss reaction has been referred to as superoxide driven Fenton chemistry (Liochev, 1996).

The requirement of H_2O_2 in the Fenton reaction led to the misleading concept of oxidative stress that ignores the fact that hydroxyl radical ($OH^{\cdot}$), known to be the most biologically active free radical, is formed *in vivo* under hypoxic conditions (Michiels, 2004) Hydroxyl radicals can be generated *in vitro* under the reducing condition in the presence of ascorbic acid and iron ions (Halliwell, 1996). Even more intriguing is a discovery that the generation of hydroxyl radicals catalyzed by ferric ions (Fe^{2+}) without any additional redox agent, which can be considered as a special case of the Fenton reaction:

$$Fe^{3+} + HO^- \rightarrow Fe^{2+} + OH^{\cdot},$$

where one electron from the hydroxyl group of water is transferred to the ferric ion with the formation of a divalent iron and a hydroxyl radical (Lipinski, 2011). Brewer reviewed the roles of both iron and copper in aging-related diseases (Brewer 2007) and concluded that normal levels of iron and copper that may be healthy during the reproductive years appear to be contributing to diseases of aging and possibly the aging process itself.

Thus the major toxicity of superoxide radical and H_2O_2 may be associated with the ability of $O_2{}^{\bullet-}$ to release Fe^{2+} by damaging 4Fe-4S cluster and initiation of the Harber-Weiss and Fenton reactions.

Oxidative damages to the iron-sulfur centers (or clusters)

Iron-sulfur centers (Fe-S) are prosthetic groups containing 2, 3, 4, or 8 iron atoms, complexed to a combination of elemental and cysteine sulfur atoms. The most vulnerable to the damage caused by superoxide radicals are 4Fe-4S clusters. The respiratory complex I has 4 such clusters, Complex II, which is also succinate dehydrogenase, has one 4Fe-4S cluster. Extreme sensitivity of the enzyme akonitase to superoxide radical is also associated with the damage to the enzyme's 4Fe-4F cluster.

The 4-Fe centers have a tetrahedral structure, with Fe and S atoms alternating as vertices of a cube, as depicted at right. The cysteine residues provide sulfur ligands to the iron, while also holding these prosthetic groups in place within the protein. Electron transfer proteins may contain multiple iron-sulfur centers as in Complex I. Iron-sulfur centers transfer only one electron even if they contain two or more iron atoms, because of the close proximity of the iron atoms. (The figures are from the website "Molecular Biochemistry").

Peroxyl radical. Lipid peroxidation

Step 1. In a peroxide-free lipid system, the initiation of a peroxidation sequence refers to the attack of a ROS (with sufficient reactivity) with abstraction of a hydrogen atom from a methylene group ($- CH_2-$) (see the Table 12.2 below).

$$-CH_2- + {}^{\bullet}OH \longrightarrow -\overset{\bullet}{C}H- + H_2O$$

This attack generates easily free radicals from polyunsaturated fatty acids. $OH^{\bullet}$ is the most efficient ROS to do that attack, whereas superoxide radical is insufficiently reactive. This peroxidation process is inhibited by tocopherols, mannitol and formate. The presence of a double bond in the fatty acid weakens the C-H bonds on the carbon atom adjacent to the double bond and so makes H removal easier. The carbon radical tends to be stabilized by a molecular rearrangement to form a conjugated diene.

Step 1.

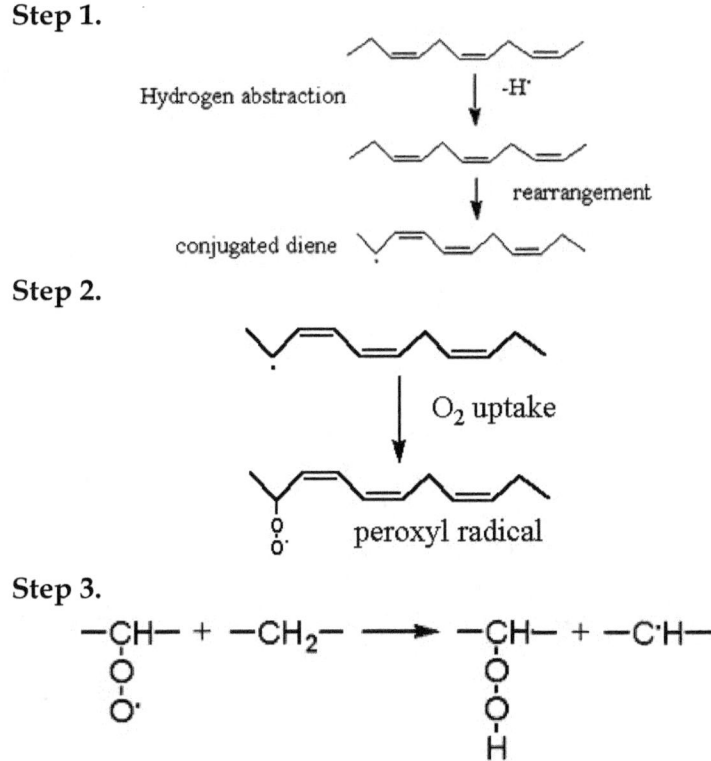

Step 2.

Step 3.

Step 2. Under aerobic conditions conjugated dienes are able to combine with O_2 to give a peroxyl (or peroxy) radical, $ROO^\bullet$.

Step 3. As a peroxyl radical is able to abstract H from another lipid molecule (adjacent fatty acid), especially in the presence of metals, such as copper or iron, thus causing an autocatalytic chain reaction. The peroxyl radical combines with H to give a lipid hydroperoxide (or peroxide). This reaction characterizes the propagation stage.

Probable alternative fates of peroxyl radicals are to be transformed into cyclic peroxides or even cyclic endoperoxides from polyunsaturated fatty acids such as arachidonic or eicosapentaenoic acids.

Lipid peroxidation is associated with formation of peroxyl radicals and some other forms of oxygen-containing radicals, such as alkoxyl radicals. Therefore it is useful to remind the reader the nomenclature of these compounds. A good definition of the terms can be found on Internet presented by Fred Senese.

Table 12.2. Alkoxyl radicals

Alkane	Formula	Alkyl group	Formula
methane	CH_4	methyl group	$-CH_3$
ethane	CH_3CH_3	ethyl group	$-CH_2CH_3$
propane	$CH_3CH_2CH_3$	n-propyl group	$-CH_2CH_2CH_3$
butane	$CH_3CH_2CH_2CH_3$	n-butyl group	$-CH_2CH_2CH_2CH_3$

Author: Fred Senese: senese@antoine.frostburg.edu

Meanings of *alk•oxy*: **1)** of, relating to, or containing a monovalent radical RO– composed of an alkyl group united with oxygen — often used in combination; **2)** any univalent radical R-O-, or anion R-O-, where R is an alkyl group.

What is an alkyl group? An alkyl group is a piece of a molecule with the general formula $C_nH_{2n}+1$, where n is some integer. For example, a methyl group (CH_3) is a fragment of a methane molecule (CH_4); n = 1 in this case. The -yl ending means "a fragment of an alkane formed by removing a hydrogen".

Peroxynitrite

Formula **[O-]ON=O; Molecular mass 62.05** (average)

The structure of the peroxynitrite anion.

Peroxynitrite is the anion with the formula ONOO–. It is an unstable structural isomer of nitrate, NO_3^-, which has the same formula but a different structure. Although peroxynitrous acid is highly reactive, its conjugate base peroxynitrite is stable in basic solution. Under laboratory conditions it is prepared by the reaction of hydrogen peroxide with nitrite: $H_2O_2 + NO_2^- \rightarrow ONOO^- + H_2O$.

In the living organisms peroxynitrite can be formed by interaction of the free radical superoxide with the free radical nitric oxide: $\bullet O_2^- + \bullet NO \rightarrow ONO_2^-$

Peroxynitrite and its conjugate acid are strong oxidants, capable of effecting one- and two electron reactions akin to those of $\bullet OH$, nitrogen dioxide (NO_2), and nitrosonium cation. Peroxynitrite is an oxidant and nitrating agent. Because of its oxidizing properties, peroxynitrite can damage a wide array of molecules in cells, including DNA and proteins.

Peroxynitrite is a particularly effective oxidant of aromatic molecules and organosulfur compounds that include free amino acids and peptide residues. Cysteine and glutathione, which are significant components of antioxidant reservoirs, are converted to disulfides. Tyrosine and tryptophan undergo one-electron oxidations to radical cations, which are competitively hydroxylated, nitrated, and dimerized. The formation of nitrotyrosine is particularly favorable, and the appearance of this product in biological samples is taken as diagnostic of exposure to peroxynitrite. Purine nucleotides are vulnerable to oxidation and to adduct formation. There are thorough reviews on formation and toxicity of peroxynitrite (Pacher et al. 2007; Szabo 2003; Szabo et al 2007;).

Nitric oxide
Molecular formula: NO; Molar mass: 30.01 g mol^{-1}

Nitric oxide, also known as nitrogen monoxide, is a molecule with chemical formula NO. It is a free radical and is an important intermediate in the chemical industry. Nitric oxide is a byproduct of combustion of substances in the air, as in automobile engines, fossil fuel power plants, and is produced naturally during the electrical discharges of lightning in thunderstorms.

Nitric oxide should not be confused with nitrous oxide (N_2O), an anaesthetic and greenhouse gas, or with nitrogen dioxide (NO_2), a brown toxic gas and a major air pollutant. However, nitric oxide is rapidly oxidized in air to nitrogen dioxide.

Reactivity. When exposed to oxygen, NO is converted into nitrogen dioxide: $2 NO + O_2 \rightarrow 2 NO_2$.

This conversion has been speculated as occurring via the ONOONO intermediate. In water, NO reacts with oxygen and water to form HNO_2 or nitrous acid. The reaction is thought to proceed via the following stoichiometry: $4 NO + O_2 + 2 H_2O \rightarrow 4 HNO_2$

NO will react with F⁻, Cl⁻, and Br⁻ to form the XNO species, known as the nitrosyl halides, such as nitrosyl chloride: $2 NO + Cl_2 \rightarrow 2 NOCl$. Nitroxyl (HNO) is the reduced form of nitric oxide

Formation in vivo. There are three isoforms of the nitric oxide synthases (NOSs) enzyme: endothelial (eNOS), neuronal (nNOS), and inducible (iNOS) - each with separate functions. NOS oxidizes the guanidine group of L-arginine in a process that consumes five electrons and results in the formation of NO with stoichiometric formation of L-citrulline. The process involves the oxidation of NADPH and the reduction of molecular oxygen. The transformation occurs at a catalytic site adjacent to a specific binding site of L-arginine. Since it takes two oxygens per NO produced, and if the half-life of NO *in vivo* was as long as 7 s, then 120 nmol O_2 would be needed per gram tissue per minute to maintain NO at a steady-state concentration of 1 μM! (Pacher et al. 2007).

The neuronal enzyme (NOS-1) and the endothelial isoform (NOS-3) are calcium-dependent and produce low levels of this gas as a cell signaling molecule. The inducible isoform (NOS-2) is calcium independent and produces large amounts of gas which can be cytotoxic.

Functions of NO. In mammals, NO is an important cellular signaling molecule involved in many physiological and pathological processes. It is a powerful vasodilator with a short half-life of a few seconds in the blood. Long-known pharmaceuticals like nitroglycerine and amyl nitrite were discovered, more than a century after their first use in medicine, to be active through the mechanism of being precursors to nitric oxide. Low levels of nitric oxide production are important in protecting organs from ischemic damage.

NO is an important regulator and mediator of numerous processes in the nervous, immune and cardiovascular systems. These include vascular smooth muscle relaxation, resulting in arterial vasodilation and increasing blood flow. NO is also a neurotransmitter and has been associated with neuronal activity and various functions like avoidance learning. NO also partially mediates macrophage cytotoxicity against microbes and tumor cells. In addition to mediating normal functions, NO is implicated in pathophysiologic states as diverse as septic shock, hypertension, stroke, and neurodegenerative diseases.

Cytotoxic Effects of Nitric Oxide. Although NO was reported to have numerous potentially toxic effects, many of them are more likely mediated by its oxidation products rather than NO itself. The most toxic product of NO oxidation is peroxynitrite, particularly at the sites were both NO and O_2^- are present. Indeed, activated macrophages produce both NO and superoxide, so the inactivation of mitochondria in tumor cells could well have been mediated by peroxynitrite (reviewed in Pacher et al. 2007).

NO may reversibly inhibit enzymes with transition metals or with free radical intermediates in their catalytic cycle. NO in micromolar concentrations will reversibly inhibit catalase and cytochrome P-450, ribonucleotide reductase, the enzyme responsible for DNA synthesis that contains a tyrosine radical. NO in the submicromolar range can also reversibly inhibit cytochrome-c oxidase (Shiva et al. 2001). Some authors suggested that this may transiently increase the leakage of superoxide from the electron transport chain (Pacher et al. 2007). These suggestions, however, to me look rather far-fetched rather than proved experimentally. First, complex IV is present in a large excess over other respiratory complexes. The redundancy of respiratory chain was well illustrated by titration with inhibitors of mitochondria oxidizing various substrates. In my experiments on titration with rotenone of Complex I, which has the lowest content in the respirasome, the changes in the State 3 respiration were observed only after approximately 80% of the enzyme was inhibited (Betarbet et al. 2000); Secondly, the transport of electrons from cytochrome c_1 to Complex IV is accompanied with huge release of energy, therefore electrons from reduced components of cytochrome oxidase cannot go back to Complex III. I have observed no increase in ROS production of rat brain mitochondria upon addition of KCN (A. Panov, unpublished data).

CHAPTER 13

Mitochondrial Calcium Transport and Permeability Transition

Introduction

In the history of Mitochondriology, calcium transport and accumulation occupies a big Chapter. Before the acceptance of the Mitchell's chemiosmotic theory, the results and ideas regarding mechanisms of Ca^{2+} transport stood apart from the main fundamental problems regarding energy transduction and oxidative phosphorylation, and thus formed a kind of a separate branch of the Mitochondriology. Initially, it was assumed that mitochondria serve as a sink for the excess of free Ca^{2+} ($[Ca^{2+}]_{Free}$) in the cells. It was discovered that after accumulation of a certain amount of Ca^{2+} and accompanying anion mitochondria undergo large amplitude swelling. Further studies have shown that swelling of the mitochondria was caused by opening of a large pore that allows free movement of solutes below 12,000 Da. As a result, sucrose and other cytosolic components were moving into mitochondria whereas mitochondrial adenine- and pyridine nucleotides were leaking into the cytosol (incubation medium). The phenomenon of the pore opening was named "mitochondrial permeability transition" (mPT), and the suggested pore was titled "mitochondrial permeability transition pore"(mPTP).

At the beginning of the 90s, many critics suggested that mPT was a blunder, an artifact, because it had no evident physiologically significant function. Then, in the middle of 90s, scientists revealed that "mitochondrial permeability transition is a central coordinating event of apoptosis" (Marchetti et al. 1996; Petit et al. 1996). This discovery had a huge impact on Mitochondriology, which at the time was in a great downturn, as well as on the relatively new branch of the cell research studying cell death - Apoptosis. Researchers exploring apoptosis have discovered dozens, if not hundreds, of death signals, peptides and proteins, receptors and so on. A very large amount of experimental data was accumulated, which at the time was almost impossible to arrange into a more or less physiologically relevant hypothesis. And suddenly, the murky fog of the data cleared, and the beautifully logical images of the mitochondrial roles in apoptotic and necrotic pathways of cell death emerged, and all the apoptotic regulation factors took their proper place. Together with the discovery at the same time of the maternally inherited mitochondrial diseases (Wallace, 1992), the discovery of the mitochondrial roles in the Life and Death of cells (Giacomello et al. 2007) resulted in the emergence of new branches of Mitochondriology – Mitochondrial Medicine and Mitochondrial Pathophysiology.

It is almost impossible to review all the twists, documented in the literature, regarding calcium transport and its regulation. Ernesto Carafoli, who was at the beginning of the events, published in 2010 a reviewing paper where he described the key events in the rather dramatic history of mitochondrial calcium handling (Carafoli, 2010). We, however, have to stay focused on the methods and methodology of mitochondrial calcium transport and permeability transition. Therefore, we will skip most of the historical events and start from the beginning, trying to answer "simple" questions: Why calcium? Why inorganic phosphate?

Calcium phosphate (CaPi) is the principal form of calcium found in the body and body's fluids. In milk it is found in higher concentrations than would be possible at the normal pH because it exists in a colloidal form in micelles bound to casein protein with magnesium, zinc and citrate, collectively referred to as colloidal calcium phosphate. Thus milk and milk products are a good source of calcium and phosphate for the body. Seventy percent of bone consists of hydroxyapatite. Tooth enamel is composed of almost 90% hydroxyapatite. In the cells, the concentrations of free calcium ions are low, and most of Ca^{2+} is unevenly distributed among different stores such as endoplasmic reticulum and mitochondria, and bound to various proteins, which have a wide range of affinities to calcium and distribution among different cell types (calmodulin, calcein, annexins, parvalbumin, calbindin-D28K, and others).

Properties of ortophosphoric acid and the calcium phosphate salts.

Of the four major cations (Na^+, K^+, Mg^{2+}, and Ca^{2+}) in the body, it is Ca^{2+} that appears to have a special messenger role in biological systems (Urry 1978). In a search for an appropriate ionic messenger it was noted that inorganic cations have a wider range of interactions with biomolecules than anions. The prevalent monovalent cations have too high flux across lipid membranes and too weak interactions with molecules in aqueous domains. The trivalent cations, on the other hand, cannot be effectively transported across lipid membranes. Thus divalent cation movement across lipid membranes can be well modulated and their divalent charge allows for a wide range of binding constants with biological molecules. For reasons of radius-compatibility with polypeptide chelation and due to the lack of stringent crystal field requirements, Ca^{2+} is a most suitable among divalent cations for a messenger role (Urry 1978). To a large degree this is also bound to the ability of calcium and inorganic phosphate to form 4 different salts that have a wide range of solubility and thus allowing precise regulation of Ca^{2+} concentrations in different tissues and cell's compartments.

The titration curve of phosphoric acid, which is also referred to as inorganic phosphate (Pi), has seven inflection points, of which three represent pKa1 = 2.12, pKa2 = 7.21, and pKa3 = 12.32 (CRC Handbook of Chemistry and Physics). Four additional points of inflection represent pH values where H_3PO_4 (at pH below 2), $H_2PO_4^-$ (at pH 4.6), HPO_4^{2-} (at pH 9.0) and PO_4^{3-} (at pH 11.9) exist as sole species (Fig. 13.1).

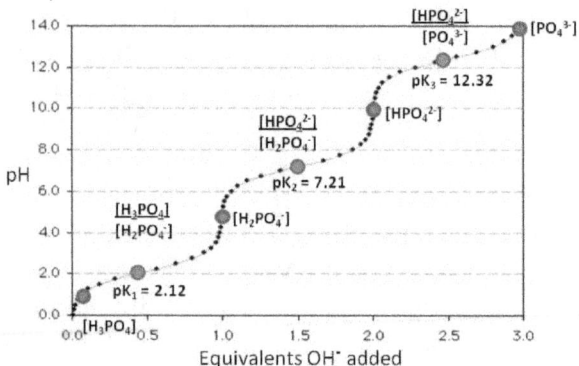

Figure 13.1 Phosphoric acid titration curve.

There are three different Ca^{2+} phosphate salts with each anion species of the orthophosphoric acid: calcium orthophosphate, mono-(prim.) $Ca(H_2PO_4)_2\ H_2O$, which is moderately soluble in water up to 18 g/L (at 30°C), calcium orthophosphate, di-(sec.) $CaHPO_4\ 2H_2O$, which is 57 times less soluble than the mono- salt, and calcium orthophosphate, tri-(tert.) $Ca_3(PO_4)_2$, which is 900 times less soluble than the mono-salt, and 15.8 times less soluble than the di- salt (Fig. 6.8.2).

$$pK_{a1} = 2.12 \qquad pK_{a2} = 7.21 \qquad pK_{a3} = 12.32$$

$$H_3PO_4 \rightleftharpoons H^+ + H_2PO_4^- \rightleftharpoons H^+ + HPO_4^{2-} \rightleftharpoons H^+ + PO_4^{3-}$$

pH 4.6   pH 9.0

$$Ca(H_2PO_4)_2 * 2H_2O \qquad CaHPO_4 * 2H_2O \qquad Ca_3(PO_4)_2$$

SOLUBILITY (g/L): 18 (100%)   0.316 (1.75%)   0.02 (0.11%)

Ca^{2+} *Orthophosphate* (mono-)   (di-)   (tri-)

Hydroxyapatite $Ca_5(PO_4)_3OH$

Figure 13.2 Types of calcium orthophosphate salts.

The forth calcium phosphate salt is hydroxyapatite, but the mechanism of its formation is poorly understood. Unlike most other compounds, the solubility level of calcium phosphates becomes lower as temperature increases. Thus heating causes precipitation.

Mechanisms of Ca^{2+} transport and accumulation

According to chemiosmotic theory (Mitchell 1977), respiration energizes mitochondria by generating the protonmotive force (Δp), which consists of two forms of energy: $\Delta\Psi$ - an electrical charge difference across the inner membrane (negative inside), and a pH gradient (more alkaline pH in the matrix), such that $\Delta p = \Delta\Psi$ –59 ΔpH. Energized mitochondria accumulate Ca^{2+} electrogenically at the expense of $\Delta\Psi$, and phosphate is accumulated at the expense of ΔpH.

The transport of phosphate occurs via specific carrier as H_3PO_4, which is equivalent to a $H_2PO_4^-$ / OH^- antiport. Therefore, when Ca^{2+} is transported into mitochondria in the presence of excess of phosphate, the H^+/Ca^{2+} ratio is close to 1.0. When mitochondria transport Ca^{2+} in the absence of anion, they extrude $2H^+$, which results in alkalinization of the matrix (Moyle & Mitchell 1977). The consumption of Ca^{2+} will stop, when ΔpH will be too large to allow further electron transport. This is the so called metabolic State 6, when Δp is represented predominantly by ΔpH. In State 6, ΔpH may be as large 3 pH units, which is equivalent to 180mV. Thus, in the absence of anions, the $H^+/Ca^{2+} \approx 2.0$. Normally, the ΔpH is close to 1 pH unit (60 mV), and $\Delta\Psi$ is equivalent to 190 mV.

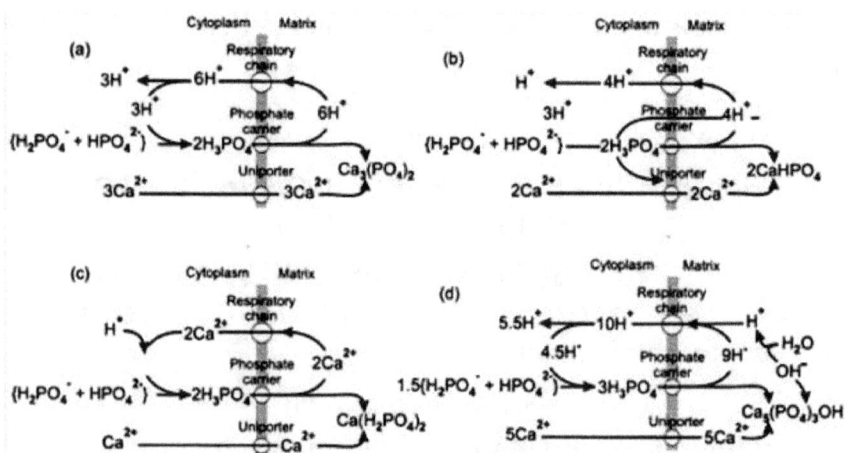

Figure 13.3. Possible stoichiometries of Ca-Pi complex formation in the mitochondrial matrix. Explanation of the figure 13.3 is in the text.

Chalmers and Nicholls (2003) provided a thorough theoretical explanation that the ratio of H^+ extruded during Ca^{2+} consumption (the H^+/Ca^{2+} ratio) reflects the type of calcium phosphate salt sequestered in the mitochondria. The interactions between Ca^{2+} and Pi in respiring mitochondria are summarized in the figure 13.3, as described by Chalmers and Nicholls (2003).

Explanation of Figure 13.3. Ion transport stoichiometry is shown for the respiration-linked mitochondrial Ca^{2+} accumulation in the presence of excess phosphate. The external pH was assumed to be 7.2, that is concentrations of $H_2PO_4^-$ and HPO_4^{2-} are equal. The phosphate carrier was assumed to function as a H_3PO_4 uniporter, this is equivalent to a $H_2PO_4^-/OH^-$ antiport. **(a)** Amorphous tricalcium phosphate, $Ca_3(PO4)_2$, forms in the matrix, a H^+/Ca^{2+} ratio of 1 is seen in the external medium; **(b)** $CaHPO_4$ formation gives a H^+/Ca^{2+} ratio of 0.5; **(c)** $Ca(HPO_4)_2$ formation gives a H^+/Ca^{2+} ratio of -1; **(d)** Hydroxyapatite formation would be associated with a H^+/Ca^{2+} ratio of 1.1. It should be noted, that under normal experimental conditions hydroxyapatite does not form in mitochondria.

Under normal conditions, pH of the cytosol is 7.2, and thus the concentrations of $H_2PO_4^{2-}$ and HPO_4^{2-} outside mitochondria are equal. Because pH in the matrix is more alkaline, the concentration of HPO_4^{2-} is higher than $H_2PO_4^{2-}$, and Ca^{2+} forms predominantly $CaHPO_4$, which has 57-fold lower solubility than calcium mono-orthophosphate (CRC Handbook of Chemistry and Physics). As a result, $CaHPO_4$ is sequestered as amorphous sediment removing equimolar amounts of Ca^{2+} and $H_2PO_4^{2-}$ from the cytosol. In the alkaline pH in the matrix, part of the ortophosphoric acid will exist as PO_4^{3-}. Chalmers and Nicholls (2003) stressed that the gradient of the PO_4^{3-} anion varies as the third power of the pH gradient across the inner membrane, thus matrix alkalization from pH 7.0 to pH 8.0 would increase the matrix PO_4^{3-} concentration by a factor of 1000. Because $Ca_3(PO_4)_2$ is practically insoluble, this salt is more preferable thermodynamically than $CaHPO_4$. However, formation of calcium tri-orthophosphate is limited kinetically by the availability of the PO_4^{3-} anion, and thus strongly depends on matrix pH.

Simultaneous consumption of $H_2PO_4^{2-}$, HPO_4^{2-} and $2Ca^{2+}$ with formation of 2 $CaHPO_4$ would result in the net extrusion of one proton and the H^+/Ca^{2+} ratio of 0.5. Formation of $Ca_3(PO4)_2$ would result in consumption of $3Ca^{2+}$ and net extrusion of 3 H^+, which results in the H^+/Ca^{2+} ratio of 1.0. Chalmers and Nicholls (2003) accepted that H^+/Ca^{2+} ratio of 1 as reported by Lehninger et al. (1967). However, they also reported H^+/Ca^{2+} ratios of 0.8 in the presence of Pi. The ratio of 0.8 is obtained when both the di- and tri-orthophosphate calcium salts are formed, when 4 net H^+ are extruded and 5 Ca^{2+} are consumed. $3Ca^{2+}$ are sequestered as $Ca_3(PO4)_2$, $2Ca^{2+}$ are sequestered as $CaHPO_4$, together with 4 molecules of H_3PO_4 (see Fig. 13.3).

Because CaPi is sequestered inside mitochondria predominantly as a quasi-insoluble salt, the mitochondrial concentration of free Ca^{2+} remains remarkably low (at about 2 μM) and constant until the opening of mPTP (Chalmers and Nicholls 2003). Thus, the constant level of $[Ca^{2+}]_{Free}$ in the matrix is the result of formation of CaPi precipitates (Chalmers & Nicholls, 2003; Greenawalt et al. 1964; Pivovarova & Andrews 2010) that can be observed in electron microscopy as dense granules of about 500 Å in diameter. The granules are always located close to the inner membrane and often show electrontransparent "cores." Such granules appear to be made up of clusters of smaller dense particles where some proteins serve as a center for CaPi precipitates (Greenawalt et al. 1964).

At this point, when we learned the principal mechanisms of calcium transport and sequestration, we turn our attention to the Methods. Without thorough assessment of the methods, it will be impossible to critically scrutinize the literature data and fully understand the roles of CaPi in the mitochondrial physiology and pathophysiology, and how to study them experimentally.

Methods to study calcium concentrations inside and outside mitochondria

In this section we will discuss the methods of Ca^{2+} determination and registration of Ca^{2+} consumption by mitochondria until opening of the permeability transition pore (mPTP), therefore it is more methodical. In the next section we will discuss how to study the phenomenon of permeability transition and its regulation. Because it relates more to mitochondrial physiology and regulation, the next section is more methodological.

In order to isolate normal mitochondria and to study transport Ca^{2+}, it is very important to prevent overloading of mitochondria with Ca^{2+} during the isolation procedure. In Chapter 1 of this book I have mentioned that too many chemicals, particularly sucrose and mannitol, are contaminated with calcium. In addition, during homogenization of the tissue, calcium is released from various stores together with Pi, and thus, mitochondria might be significantly overloaded with CaPi, if there was no chelator present in the isolation medium. For example, rat liver mitochondria isolated in the presence of 1 mM EGTA contained 3.8 ± 0.4 nmol Ca^{2+} per mg RLM, whereas mitochondria isolated without the chelator contained 18.4 ± 0.2 nmol Ca^{2+} per mg RLM. EGTA has little effect on the mitochondrial Mg^{2+} contents, which in the above experiments were correspondingly 16.3 ± 0.8 and 22.5 ± 0.7 nmol Mg^{2+} per mg RLM (Panov & Scarpa, 1996). In the cell, the cytosolic $[Ca^{2+}]_{Free}$ is about 100 nM (Somlyo et al., 1985), and the mitochondrial $[Ca^{2+}]_{Free}$ (as determined by Fura 2AM) is less than 1 µM (A. Panov). Somlyo, using the electron probe method, have shown that mitochondrial $[Ca^{2+}]_{Free}$ content was about 0.3 µM based upon a total Ca^{2+} content of 1.1 nmol per mg RLM (Somlyo et al. 1985). Thus our results and those of Somlyo are in a good agreement taking into consideration differences in the methods and conditions.

It should be mentioned that there exist many more methods to quantify the amounts of Ca^{2+} in media and mitochondria, than described in this section. I present here only those methods, with which I have personal experience and thus can describe the pitfalls and advantages of the methods.

Determination of the total Ca^{2+} and Mg^{2+} by Atomic Absorption Spectroscopy.

I used the method described by McDonald et al. (1976). This method allows measuring the total mitochondrial contents of Ca^{2+} and Mg^{2+}. At the Case Western University I used the Perkin-Elmer absorption spectrometer. Because calcium can be extracted from glass, all procedures were performed using 1.5 ml plastic tubes. The isolated mitochondria were suspended in 0.25 M sucrose buffered with 10mM MOPS, pH 7.2. The aliquots (1–2 mg) of each sample of the mitochondria were placed in 2-3 plastic tubes for parallel measurements. The mitochondria were sedimented at high speed using a table microcentrifuge. The sediments were then solubilized with 0.1 ml 10 M NaOH using small Teflon pestle, and then heated to 75°C in a water bath for approximately 5 min. Following this, 1 ml of the solution containing 0.01 M lanthanum oxide, 0.025 M HCl, 0.02 M EDTA, and 0.08 M NaOH was added. Samples were again heated to 75°C, for approximately 5 min, and then analyzed by atomic absorption spectrophotometry. It should be mentioned that there are methods with somewhat different treatments of the samples.

Determination of the mitochondrial $[Ca^{2+}]_{Free}$ using Fura-2 AM.

Fura-2 is a calcium imaging dye that binds to free Ca^{2+}. Because this indicator has very high affinity to Ca^{2+}, it is practically unsuitable for determination of calcium retention capacity. Fura-2 AM is the cell-permeable acetoxymethyl (AM) ester form of Fura-2. The excess of Fura-2 AM is added to energized mitochondria and incubated for 15-20 minutes. During this time most of the acetoxymethyl ester is hydrolyzed, and Fura-2 becomes trapped inside mitochondria. After that, mitochondria are sedimented, washed once with a medium and sedimented again. These mitochondria are used for measurement of the $[Ca^{2+}]_{Free}$ in the matrix. In order to maintain low $[Ca^{2+}]_{Free}$ outside mitochondria, similar to the cytosol, it is recommended to utilize the Ca^{2+}/EGTA buffer. At the Ca^{2+}/EGTA of 0.7 (0.7 mM $CaCl_2$ and 1 mM EGTA) the $[Ca^{2+}]_{Free}$ in the matrix of mitochondria is less than 1 µM as determined by the fura-2 AM (A. Panov).

Fura 2: Molecular weight: 1001.86. Peak excitation: variable depending on the concentration of free Ca^{2+}, between 300 and 400 nm; **Peak emission:** 510 nm. Fura-2 AM should be diluted in high-quality, freshly opened DMSO. Once diluted, it should be protected from light and stored at -20°C. Avoid freeze-thawing. Upon binding Ca^{2+}, the excitation spectrum of Fura-2 shifts to shorter wavelengths between 300 and 400 nm, while the peak emission remains steady around 510 nm. The K_d of Fura-2 is highly dependent on pH, temperature, ionic strength and viscosity of the cytosol, thus great care should be taken when non-standard conditions are used.

Methods of Ca^{2+} determination outside mitochondria by fluorescent indicators.

The Calcium Green™-5N method. Calcium Green-5N is available as a cell-impermeant potassium salt (Invitrogen catalog number C3737), or as a cell-permeant AM ester (C3739). Calcium Green™-5N exhibits an increase in fluorescence emission intensity upon binding Ca^{2+} with little shift in wavelength. Calcium Green™-5N is essentially nonfluorescent in the absence of Ca^{2+}. It is a relatively low-affinity indicator with the dissociation constant for Ca^{2+} in the absence of Mg^{2+} of about 14 μM. **Exitation** 506 nm/**Emission** 532 nm.

There is available a wide selection of similar indicators with different affinities to Ca^{2+}. For the purpose of following Ca^{2+} consumption by mitochondria it is better to use an indicator with lower affinity for the cation.

Cell-impermeable Calcium Green-5N and Magnesium Green are perfect for the purpose. These indicators buffer intracellular Ca^{2+} to a lesser extent than do the higher-affinity Ca^{2+} indicators. Furthermore, the high Ca^{2+} dissociation rates of these indicators are advantageous for tracking rapid Ca^{2+}-release kinetics. Because Calcium Green-5N has relatively lower affinity, it is better to use than Magnesium Green to follow Ca^{2+} consumption added as a sequence of aliquots to study calcium retention capacity (see Fig. 13.4).

Figure 13.4A shows that unprotected mouse liver mitochondria (MLM) underwent permeability transition just after the 3d addition of the calcium aliquot containing 50 nmol Ca^{2+}/mg mitochondrial protein. MLM protected with cyclosporine A + ADP opened the mPTP upon addition of 6 aliquots of Ca^{2+} (Fig. 13.4B). Thus unprotected MLM sequestered about 120 nmol Ca^{2+} per 1 mg, and protected MLM sequestered 340 nmol Ca^{2+} per 1 mg of mitochondrial protein. We designate the amount of Ca^{2+} sequestered by mitochondria before opening of mPTP as the calcium retention capacity.

Calcium Retention Capacity (CRC).

This is a quantitative parameter, which allows comparison of mitochondria from different organs, or species. Thus, using Calcium Green-5N indicator, you can quantitatively evaluate mitochondrial capacity to accumulate CaPi.

From the Fig. 13.4 one can see that just before addition of mitochondria there was a significant fluorescence of the indicator, which suggests that the incubation medium contained about 25 nmol Ca^{2+}/ml. This is how I discovered that the chemicals were contaminated with calcium. The background shown in the Fig. 13.4 was observed after I took all precautions by switching to chemicals of the highest quality and isolating mitochondria, including the final centrifugation, in the presence of 1 mM EGTA.

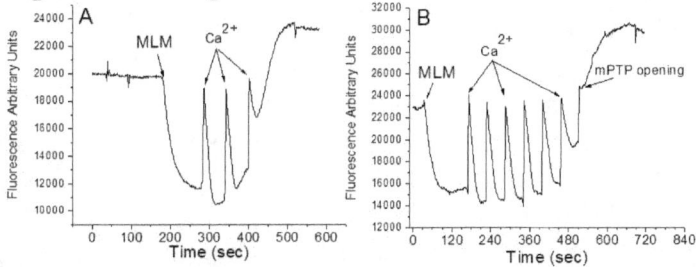

Figure 13.4. Determination of the calcium retention capacity by Calcium Green-5N. A. Unprotected mouse liver mitochondria; B. Mouse liver mitochondria protected by oligomycin 5 μg, cyclosporin A 0.5 μM, and ADP 25 μM. Incubation conditions: KCl 120 mM. NaCl 10 mM. $MgCl_2$ 0.5 mM. KH_2PO_4 1 mM. Glycyl-Glycin 3 mM, pH 7.2, glutamate 10 mM. malate 2 mM, Calcium Green-5N 25 nM. Final volume 2 ml, 1.0 mg rat liver mitochondria from a FVB/N mouse. Calcium additions: aliquots of 25 nmol/ml (5 μl of 10 mM stock $CaCl_2$).

Like many other researchers, I performed the last centrifugation in the medium without EGTA, in order to avoid the presence of the chelator in the working suspension of mitochondria. In those experiments, addition of mitochondria to the medium containing Calcium Green-5N resulted in the burst of the fluorescence, which made any measurements of Ca^{2+} impossible.

If the final sedimentation of mitochondria was performed in the medium with 1 mM EGTA, the working suspension of mitochondria in the 0.25 M sucrose without EGTA still would contain some amount of the chelator. My evaluations suggested that the working suspensions of mitochondria would contain about 25-40 μM EGTA depending on the volume of the sucrose used for dilution of mitochondria. That is why addition of mitochondria to the incubation medium resulted in a sharp drop of fluorescence to zero level (Fig. 13.4). This small amount of EGTA evidently did not preclude measurements of calcium transport by mitochondria.

Determination of the extramitochondrial Ca^{2+} with the Ca^{2+}-sensitive electrode

Table 13.1. Ingredients for manufacturing the Ca^{2+}-sensitive electrode.

Ingredients	Petri dish D = 48 mm
PVC (high Mol. Wt) Polyvinylchloride resin. This amount of PVC, which makes right thickness of the membrane (30.9% of all components).	72 mg
Tetraphenylboron (borate) $NaBPh_4$ (2.1%) Fluka.	1.25 mg
ETH 469 (Bis(1-butylpentyl)decane-1,10-diyl diglutarate) Plasticizer of high lipophilicity (63.7%).	131.4 µl
ETH1001 (Ca^{2+} ionophore I, This is a neutral ligand (-)-(R,R)-N,N'-Bis-[11-(ethoxycarbonyl)undecyl]-N,N'-4,5-tetramethyl-3,6-dioxaoctane-diamide (3.3%). Fluka	7.7 mg
Tetrahydrofuran (THF).	2-3 ml

Preparation of the Ca^{2+}-sensitive electrode was described by Tsien & Rink (1980). For the Petri dish of a diameter of 48 mm or 88 mm mix the ingredients as shown in Table 13.1.

Commentary 1._Tetraphenylboron (TPB) $NaBPh_4$ sold by Sigma may be of a very bad quality (at least during the period of 1990 -2010). The membrane made with Sigma's TPB would work just for few days. I used TPB from Aldrich, and the electrodes remained sensitive for months.

Commentary 2. In the original paper describing the manufacture of the PVC based electrodes, the recommended volume of THF was 480 and 960 µl for the dishes with D = 24 mm and D = 88 mm. It is more practical to use much more THF for better dissolving and even distribution of the mixture in the dish. Cover the dish and THF will slowly evaporate.

Commentary 3. THF is very TOXIC! Work in a hood, and store the freshly-made membrane under hood for 2-3 days and then between the two sheets of waxed paper between pages of a book.

Commentary 4. Because the Ca^{2+}-sensitive electrode has PVC membrane with Tetraphenylboron, it will respond to TPP^+ and similar compounds. Therefore, you cannot simultaneously utilize Ca^{2+} electrode and TPP^+ electrode.

Dissolve the components in THF by using magnetic stirrer and do not shake. The little glass vessel in which you prepare the THF solution must be tightly closed. Open it only for addition of more THF. You may add more THF (the final volume is not very essential), and after the components become completely dissolved (the solution must be transparent), pour the solution into the Petri dish and leave covered with the lid in the hood to evaporate slowly. After drying (usually next day), the membrane will be soft and easily detach from the Petri dish. Using a sharp small scalpel detach the membrane around the perimeter of the Petri dish, and with the help of two small pincers carefully remove the membrane from the dish. Place the membrane between the two sheets of thick waxed paper. Store under pressure (inside a book) to keep it evenly flat. The membrane can be used for years.

Assembling the Ca^{2+}-sensitive electrode. For the electrode you can use glass disposable microsampling 200 µl pipettes. After cutting the glass pipette of suitable length, dull the sharp end of the tube with an abrasive or over the flame of the gas burner. Otherwise the tight Tygon tube will be difficult to fit. Insert the glass capillary into a piece of the Tygon tube (15-20 mm long). The working end of the Tygon tube, to which the membrane will be glued, must be even and smooth. For this, smear a small drop of THF on the glass surface (the cover of the Petri dish will do well) and rub the end of the Tygon tube over wet surface of the glass.

Further procedures of gluing the membrane to the Tygon tube are the same as for manufacturing of the TPP^+-electrode described in Chapter 11. Inside the Ca^{2+} electrode place the solution of $CaCl_2$ in 100 mM KCl. The actual amount of $CaCl_2$ depends on the expected measured Ca^{2+} concentration. It should be 100-1000 fold higher than the measured $[Ca^{2+}]$ outside. Thus, if you plan to use additions of Ca^{2+} in 25 µM aliquots (final concentration) during Ca titrations of mitochondria, and expect that all Ca^{2+} will be released during permeability transition after, say 20 Ca additions. In that case, the total final [Ca^{2+}] would be 0.5 mM. Thus, inside the electrode the 100-fold higher $CaCl_2$ solution should be at least 50 mM. When the electrode will not be in use for several weeks, store the electrode dry. When not in use for few days, store the electrode immersed into incubation buffer.

Table 13.2. Ca^{2+}-ligands and buffers for solutions with different pCa.

pCa	[Ca] mM	Ca^{2+} -ligand	[KCl] mM	pH buffer	pH
3	1	none	98	Mops	7.30
4	5	NTA	90	Hepes	7.39
5	5	NTA	90	Taps	8.42
6	5	HEEDTA	90	Hepes	7.70
7	5	EGTA	90	Mops	7.80
8	5	EGTA	90	Hepes	7.8
zero	0	EGTA	100	Hepes	7.8

Abbreviations for Ca^{2+} ligands: NTA = Nitrilotriacetic acid; HEEDTA = N-(2-hydroxyethyl) ethylenediamine-N,N',N'-triacetic acid; Taps = N-tris(hydoxymethyl) methyl-3-aminopropanesulfonic acid.

Commentary. First, prepare 100 ml of the stock solution of $CaCl_2$. Use ultrapure $CaCl_2$ from Sigma and all solutions are prepared using precise measuring flasks and deionized water. Adjust pH with KOH.

For calibration of the Ca^{2+}-sensitive electrode prepare a number of solutions with exact pCa as described in the paper by Tsien & Rink (1980). Each solution with a given pCa contains 10 mM of the specified pH buffer and 10 mM of a Ca^{2+} ligand, with exception of the solution with pCa 3, which contains no ligand. The pH of each solution should be brought to the specified level by titration with KOH. The necessary Ca^{2+}ligands and the buffers are listed in Table 6.8.2.

Related references:

Caroni, P., Gazzotti, P., Vuillemier, P., Simon, W., Carafoli, E. (1977) Ca^{2+} transport mediated by a synthetic neutral Ca^{2+}-ionophore in biological membranes. Biochim. Biophys. Acta. 470, 437-445.

McGuigan J.A.S.,Lutji D.,Buri A. (1991) Calcium buffer solutions and how to make them: A do it yourself guide Canad. J. Physiol. Pharmacol. 69, 1733-1749.

Simon, W., D. Ammann, M. Oehme, W.E. Morf. (1878) Calcium-selective electrodes. Ann. N.Y. Acad. Sci. 307, 52-69.

Simon, W., E. Carafoli. (1979) Design, properties and applications of neutral ionophores. Methods in Enzymology. 56, 439-448.

Tsien, R.Y., T.J. Rink. (1980) Neutral carrier ion-selective microelectrodes for measurement of intracellular free calcium. Biochim. Biophys. Acta. 599, 623-638.

Methods of registration of the mitochondrial permeability transition; calculation of the H^+/Ca^{2+} ratio and estimation of the calcium retention capacity (CRC).

Mitochondrial swelling as the method for registration of the mitochondrial permeability transition. Historically, it happened that for a long time calcium consumption by mitochondria and the phenomenon of permeability transition were studied on rat and mouse liver mitochondria. Even to this day, some researchers believe that there is no principal difference between liver mitochondria and mitochondria from other organs. We have already discussed in Chapter 6 that in fact, by many properties liver mitochondria stand apart from mitochondria of other organs.

The uniqueness of the liver mitochondria reflects the distinctive features of the liver's specific functions and metabolism. Liver mitochondria hold within the matrix space a large number of enzymes that are not associated with the inner membrane, and metabolites from various metabolic pathways, which are absent in mitochondria from other organs. Therefore, by necessity the matrix of liver mitochondria possesses a relatively large osmotically active space as compared with the heart or brain mitochondria. In comparison with the cytosol, the matrix space of mitochondria contains higher concentration of KCl, substrates and higher oncotic pressure, which is osmotic pressure, created by proteins. Therefore upon permeability transition liver mitochondria undergo the so called large amplitude swelling.

Figure 13.5 shows that after several additions of Ca^{2+} aliquots mouse liver mitochondria underwent permeability transition (mPT) that resulted in a rapid decrease of the optical density by 0.4 units of OD, which was further decreased by 0.4 units of OD after addition of alamethicin. Alamethicin is a 20 amino acid peptide antibiotic, produced by the fungus *Trichoderma viride* (different isoforms of alamethicin are currently produced synthetically). In cell membranes, alamethicin forms voltage-dependent ion channels by aggregation of four to six molecules (Hall et al. 1984; Woolley & Wallace 1992).

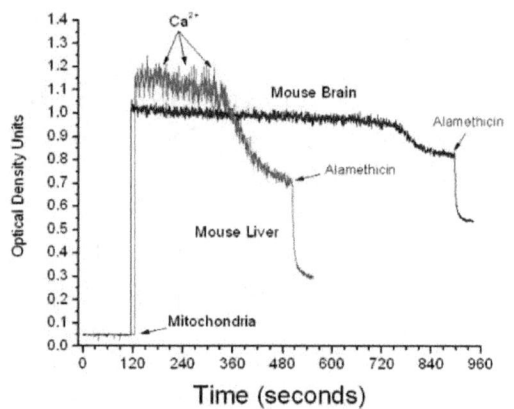

Time (seconds)

Figure 13.5. Large amplitude swelling of mouse liver and brain mitochondria upon opening of the Ca^{2+}-dependent mitochondrial permeability transition pore. Incubation conditions: KCl 120 mM, NaCl 10 mM, $MgCl_2$ 0.5 mM, KH_2PO_4 1 mM, glycyl-glycine 3 mM, pH 7.2, glutamate 10 mM, malate 2 mM, mitochondria 0.5 mg/ml, volume 2 ml. **Additions**: Ca^{2+} 12.5 nmol/ml, alamethicin 5 μg/ml. Swelling of mitochondria was registered as a decrease in optical density at 545 nm using Shimadzu spectrophotometer Multispec model 1501 (Panov et al. 2007).

The mouse brain mitochondria require significantly more additions of Ca^{2+} to initiate mPT, and the Ca^{2+}-induced decrease of the OD was 2-fold lower than in MLM (Fig. 13.5). This is because MBM have significantly higher rates of respiration and thus higher H^+/Ca^{2+} ratios. As a result, several times more CaPi was sequestered as a quasi-insoluble $Ca_3(PO_4)_2$, and thus more calcium additions were required to reach the critical $[Ca^{2+}]_{Free}$ to initiate mPT. I have intentionally did not used arrows to indicate the moments of calcium additions in order to stress that the swelling method itself shows no response to additions of Ca^{2+}. The matrix of brain mitochondria has much smaller the osmotically active space as compared to liver mitochondria. That is why the total decrease of the OD cause in the presence of alamethicin was 0.8 OD for MLM and only 0.4 OD for the MBM (Fig. 13.5).

To my experience, the amplitude of the Ca^{2+}-induced swelling is subject to wide variations that depend on the organ of origin of the mitochondria, and the animal's phenotype. In some phenotypes of rats, brain mitochondria do not undergo large amplitude swelling even after mPT. Isolated mitochondria from the PC-3 prostate cancer cells consumed large amounts of CaPi without opening spontaneously mPTP even after addition of alamethicin (Panov & Urenbayeva, 2013).

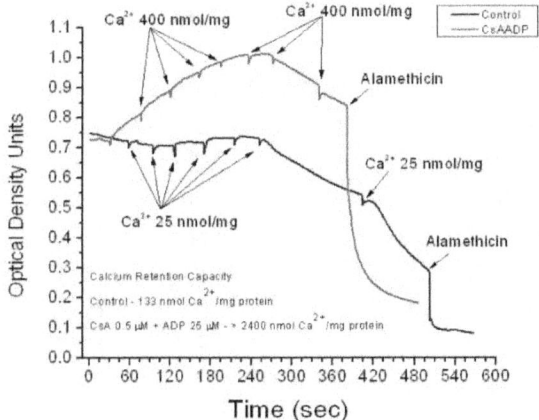

Figure 13.6. Swelling of mouse skeletal muscle mitochondria induced by calcium phosphate. Incubation conditions as in Fig. 13.5.

Figure 13.6 illustrates another example of changes in the optical density during titration of the mouse skeletal muscle mitochondria with calcium. Registration of the mitochondrial permeability transition by following the mitochondrial swelling has two major drawbacks. First, the method is purely qualitative, as can be seen from the figure 13.5. Usually the researcher does not see responses from optical density upon addition of a Ca^{2+} aliquot before mPT. In the experiment presented on Fig. 13.6, in order to see the moment of Ca^{2+} addition on the trace of optical density, the tip of the pipetter was intentionally inserted right into the light beam. This caused a very large spike, which later had to be corrected manually on the spreadsheet chart. This is a time consuming procedure and not always can be performed. The second drawback of the swelling technique is that not all mitochondria undergo large amplitude swelling. The examples are brain mitochondria and mitochondria from the cultured cells. Therefore, this method is not fully reliable and informative for studying mitochondrial mPT and for estimation of the total amount of CaPi sequestered by mitochondria.

Figure 13.6 also illustrates that in skeletal muscle mitochondria protected with CsA + ADP sequestration of CaPi caused a significant increase in the optical density of the mitochondria due to formation of the dense sediment of $Ca_3(PO_4)_2$. As a result, the protected mitochondria sequestered almost 20 times more CaPi than the control mitochondria. Unlike brain mitochondria, skeletal mitochondria do undergo large amplitude swelling upon the Ca^{2+}-induced mPT.

Table 13.3. Composition of the medium for registration of optical density registration of mitochondrial suspension.

Components	MW	500 ml
120 mM KCl	74.56	4g 474 mg
10 mM NaCl	58.44	292 mg
0.5 mM $MgCl_2$ $6H_2O$	203.3	50.8 mg
1 mM KH_2PO_4	136.09	68.0mg
K_2HPO_4	174.18	87.1 mg
Glycyl-glycine 3 mM pH 7.2	132.12	198.18 mg
Or 10 mM MOPS	209.3	1g 047 mg

The method to study changes in the optical density of mitochondria. As was mentioned earlier, most of research on the mitochondrial Ca^{2+} transport and permeability transition was conducted on liver mitochondria. The researchers usually used the sucrose medium (0.25 M) to study mPT by following mitochondrial swelling. As a substrate for initiation of the energy-dependent Ca^{2+} transport, the researchers very often used succinate + rotenone. From our current knowledge of the mitochondrial functions, the usage of rotenone is unacceptable because the inhibitor alters normal relationships between the respiratory complexes and the redox events in mitochondria. In addition, what is acceptable for the liver mitochondria is absolutely unacceptable for mitochondria from other organs. Therefore, the sucrose medium can be used for studying the mPT only if you have a special task. In most experiments, the composition of the medium and substrate selection must be as close to physiology of the mitochondria as possible.

Mitochondria from brain, heart and kidney have significantly lower rates of respiration in the sucrose medium as compared with the KCl-based media. For the maximum rates of respiration these mitochondria require K^+, and are very sensitive to the absence of Mg. In addition, NaCl is a natural component for these mitochondria. The presence of 1-2 mM Pi is optimal because at these concentrations the H^+/Ca^{2+} ratios are more controllable than at higher concentrations of Pi (Chalmers & Nicholls 2003; Panov et al. (2004). Therefore, for experiments on the mitochondrial mPT, I recommend the composition of the incubation medium, shown in Table 13.3, which worked well with all types of mitochondria I have studied.

This medium worked perfectly well also with the liver mitochondria. It should be noted that in contrast to brain, heart and kidney mitochondria, the liver mitochondria are in general much less sensitive to the changes in composition of the incubation media. Although Mg^{2+} is known to be a competitive inhibitor of the electrogenic Ca^{2+} transport, it is more correct, from the physiological point of view, to study CRC and mPT in the presence of 0.5 mM $MgCl_2$. This is because Mg^{2+} ions always present in the cytosol, and oxidation of pyruvate and α-ketoglutarate in the brain and heart mitochondria are very sensitive to the absence of Mg^{2+}. The presence of 3 mM Glycyl-glycine as a buffer, instead of MOPS, was necessary because I have usually simultaneously used the pH method to study the mitochondrial Ca^{2+} transport.

The selection of substrates depends on the type of mitochondria. The optimum substrates mixtures for the brain mitochondria are 10 mM glutamate + 2.5 mM pyruvate + 2 mM malate; for the heart mitochondria – 2.5 mM pyruvate + 5 mM succinate + 2 mM malate, or 25 μM palmitoyl-carnitine + 2.5 mM pyruvate (or 5 mM succinate). For the liver mitochondria the maximum rate of respiration can be obtained with 5 mM succinate + 5 mM glutamate + 2 mM malate without the presence of rotenone.

Spectrophotometric measurement of optical density of the mitochondrial suspension. Classical measurements of the mitochondrial swelling are performed at 520 nm using the split beam setting for spectrophotometer. However, because mitochondria are turbid, it is practical to increase the sensitivity by using the saturated solution of $CuSO_4$ as a reference solution. The highest compensation of the mitochondrial turbidity by the solution of $CuSO_4$ occurs at 545 nm. This method allows measuring swelling of mitochondria at concentrations up to 0.5 mg/ml.

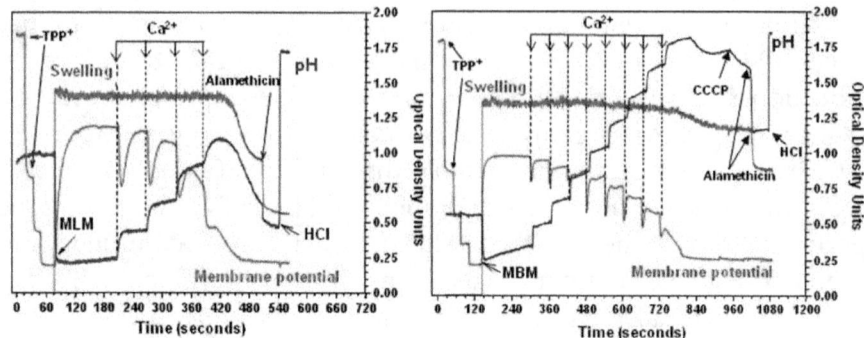

Figure 13.7. Three methods of registration of the mitochondrial permeability transition of mouse liver (MLM) and brain mitochondria (MBM) during titration with calcium. Optical density changes at 545 nm (swelling of mitochondria); Membrane potential registered with a TPP+ electrode; Changes in pH of the medium upon addition of the Ca^{2+} aliquots. **Incubation conditions**: as in Fig. 13.5. Additions: Mitochondria 0.5 mg/ml, TPP+ 0.5 µM, Final concentration 1.5 µM for MLM, 2.0 µM for MBM, alamethicin 5 µg/ml, CCCP 0.5 µM, Ca^{2+} 12.5 nmol/ml aliquots, addition of 125 nmol/ml HCl caused ΔpH of 0.07.

Mitochondrial membrane potential as the method for registration of the mitochondrial permeability transition

Figure 13.7 depicts simultaneous registration of the Ca^{2+}-induced mitochondrial permeability by three different methods: changes in the mitochondrial optical density (swelling of mitochondria), by following the changes in the mitochondrial membrane potential, and by following the changes in the pH of the incubation medium.

As we can see, the swelling of mitochondria register only the moment of mPTP opening, but provides no information about the amount of Ca^{2+} sequestered by mitochondria before mPT. The membrane potential traces, however, do provide information about how many Ca^{2+} aliquots were added to mitochondria before mPT. However, the flaw of this method is that the opening of the permeability transition pore is always accompanied by collapse of the membrane potential, but collapse of the membrane potential is not always accompanied by permeability transition. This situation is illustrated by experiment presented in Figure 13.8.

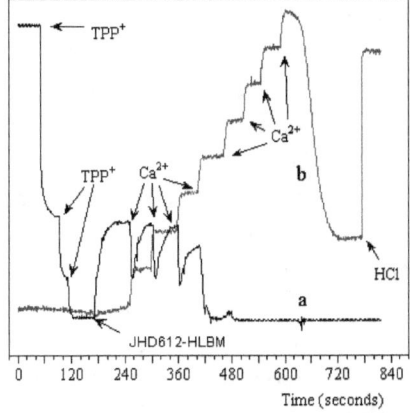

Time (seconds)

Figure 13.8. Response to calcium loading of mitochondria isolated from the human lymphoblastoid cells derived from a patient with juvenile Huntington's disease. Trace "a" – Membrane potential registered with a TPP+ electrode; **trace "b"** – changes in pH of the medium. Additions: Mitochondria 0.5 mg/ml. Final concentration of TPP+ was 1.5 μM. Downward deflection of the TPP+ trace corresponds to a decrease in membrane potential. Ca²+ aliquots were 25 nmol/mg protein. Downward deflection of the pH trace shows beginning of alkalization that reflects the moment of a large conductance pore opening. HCl was added for calibration equivalent to 62.5 nmol H+/ml, ΔpH = 0.07. (From Panov et al. 2005).

When I showed this experiment to David Nicholls at the Neuroscience conference, at first, David rejected the possibility that after the mitochondrial membrane potential collapsed, the mitochondria would continue to transport and sequester CaPi. The results of our discussions, which we had for a number of years during scientific conferences, were our papers: Chalmers & Nicholls (2003), and Panov et al. (2004).

Based on our discussions and the papers mentioned above, the explanation of this indeed remarkable experiment is as follows. CaPi is sequestered in the matrix as $Ca_3(PO_4)_2$, which has a very low solubility. Thus, Ca^{2+} consumption and sequestration of CaPi by mitochondria can be split in two steps: The first step comprises the electrogenic transport of Ca^{2+} towards the negative charge of the inner membrane, thus using the $\Delta\Psi$, and transport of H_3PO_4 towards the more alkaline pH of the matrix space for the expense of ΔpH. The second step is the process of sequestration of CaPi, which strongly depends on pH in the matrix.

As was demonstrated by Chalmers & Nicholls (2003), the mitochondrial $[Ca^{2+}]_{Free}$ remains low until the opening of mPTP. In spite of the fact that $\Delta\Psi$ constitutes the major part in mV (about 180-190 mV) of the overall electrochemical potential across the inner membrane ($\Delta p = \Delta\Psi$ -59 ΔpH, with $\Delta p_{Max} = 250$ mV)), the $\Delta\Psi$ has very low effective capacitance. On the other hand, the matrix pH is highly buffered, and therefore remains remarkably stable. Evidently, the specific properties of the lymphoblastoid cell mitochondria of patients with Huntington's disease are such, that during consumption and sequestration of CaPi, the Δp shifts to increased ΔpH.

In addition, the properties of the TPP$^+$ electrode do not allow reliable determination of $\Delta\Psi$ when there is a competition between the electrogenic diffusion TPP$^+$ molecule and very fast electrogenic transport of Ca^{2+}.

I also suggest that during the massive sequestration of CaPi in mitochondria, as shown for the brain mitochondria in Fig. 13.7, the downward deflection of the TPP$^+$ trace reflects not only the possible decrease in $\Delta\Psi$, but also ejection of TPP$^+$ by the accumulated calcium triphosphate, which occupies the scarce volume of the matrix space. I have already mentioned that mitochondrioa isolatred from the cultured cells do not undergo swelling even after addition of alamethicin. This presumes that mitochondria from cultured cells have very small osmotically actiove matrix space.

From the experiments presented above, we can conclude that measuring the mitochondrial membrane potential to study calcium transport and the mPTP is also not the optimal option in some experiments.

Method of evaluation of Ca^{2+} transport by measurement of the mitochondrial membrane potential.
The method was described earlier in Chapter 11. Here I have to Commentary the following technical points of application of this method when studying Ca^{2+} transport:

Commentary 1. During simultaneous measurements of $\Delta\Psi$ with a TPP$^+$ electrode and changes in pH in the medium with a glass pH electrode, you have to remember that because both methods require the Ag/AgCl reference electrode, be sure that both the TPP$^+$ and the glass pH electrodes have separate reference electrodes (placed in separate tubes), each connected to the incubation chamber by a separate agar-agar-KCl bridge.

Commentary 2. You cannot simultaneously measure membrane potential with a TPP^+ electrode, and Ca^{2+} with the Ca^{2+}-sensitive electrode, or measure pH with the electrode that also has the pH-sensitive PVC membrane. This is because both the calcium and the pH electrodes will respond to additions of TPP^+.

Measurements of pH changes in the medium as the method for registration of the mitochondrial permeability transition.

Of all methods I used to study the Ca^{2+}-dependent permeability transition, measurements of pH changes in response to addition of Ca^{2+} aliquots are the most reliable and informative (see Figures 13.7 and 13.8). The pH method allows not only to register every addition of a Ca^{2+} aliquot, and thus quantitatively evaluate the total amount of CaPi sequestered by mitochondria (CRC), but also to calculate the H^+/Ca^{2+} ratio, and thus estimate the type of CaPi salt sedimented in the matrix. For the experiments presented in Fig. 13.7 we calculated that the average H^+/Ca^{2+} ratio for the MLM was 0.81 ± 0.01, while for the RBM the ratio was 0.93 (Panov et al. 2007). The ratio of 0.8 is obtained when both the di- and tri-orthophosphate calcium salts are formed, when 4 net H^+ are extruded and 5 Ca^{2+} are consumed. $3Ca^{2+}$ are sequestered as $Ca_3(PO4)_2$, $2Ca^{2+}$ are sequestered as $CaHPO_4$, together with 4 molecules of H_3PO_4. The H^+/Ca^{2+} ratio higher than 0.8 means that more calcium phosphate is sequestered as $Ca_3(PO_4)_2$, which is much less soluble than $CaHPO_4$.

Moreover, the pH trace allows not only determining the calcium retention capacity of the mitochondria, but also evaluating the behavior of the permeability transition pore. This can be illustrated by comparing the pH traces (blue lines) for the MLM and MBM on Fig. 13.7. The pH trace for the MLM shows that after the 4th addition of the Ca^{2+} aliquot (25 nmol/mg), the mPTP opens and the medium's pH begins to become more alkaline. The pH trace for MBM shows that after opening of the mPTP, the slow alkalization stopped, and MBM started to consume Ca^{2+} again (Fig. 13.7). The pore opening was stimulated by addition of uncoupler (CCCP), but again the alkalization was slow, and could be stimulated by addition of alamethicin. This agrees with the high H^+/Ca^{2+} ratio (0.93) in MBM and sequestration of CaPi as insoluble $Ca_3(PO_4)_2$; therefore, upon de-energization the concentration of $[Ca^{2+}]_{Free}$ in the matrix could not raise fast enough to reach the critical level and open the large pore.

The H^+/Ca^{2+} ratios strongly correlate with the rate of the State 3 respiration, which reflects the capacity of the respiratory chain to restore $\Delta p=\Delta\Psi$ -59 ΔpH. Mouse brain mitochondria have significantly higher rate of oxidative phosphorylation (State 3 respiration) than mouse liver mitochondria, and thus, with the H^+/Ca^{2+} ratio of 0.93 MBM the calcium retention capacity (CRC) was 2-fold higher than the CRC for the MLM (Figure 13.7). In other studies we also found that even relatively small changes in the H^+/Ca^{2+} ratio result in dramatic changes of CRC. Therefore, when studying mitochondria from diseased animals, it is much more informative to study both respiratory activity and permeability transition by pH method.

The pH measuring technique is based on the classic publication of Mitchell & Moyle: Mitchell P. & Moyle J. (1967) Acid-base titration of rat liver mitochondria. Biochem. J. 104, 588-600. The potentiometric measurements of pH changes of the incubation medium during Ca^{2+} accumulation and release by mitochondria were described in Panov et al. (2004, 2007). The H^+-sensitive glass mono pH microelectrode from Lazar Co. and the Ag/AgCl-cell reference system were connected to a pH meter from Corning, model 440. It is important that the pH meter has an output for recorder. The recorder output of the pH meter was connected to the 2 channel Amplifier. The channel for the pH signal was set for 100-fold amplification, and the output from the amplifier was connected to the PC desktop computer for Data Acquisition and to the two-channel recorder.

Table 13.4. Composition of the medium for registration of pH changes during consumption and release of Ca^{2+} by mitochondria.

Components	M W	5 00 ml
120 mM KCl	74.56	4g 474 mg
10 mM NaCl	58.44	292 mg
0.5 mM $MgCl_2$ $6H_2O$	203.3	50.8 mg
1 mM KH_2PO_4	136.09	68.0mg
K_2HPO_4	174.18	87.1 mg
Glycyl-glycine 3 mM pH 7.2	132.12	198.18 mg

According to Mitchell & Moyle (1967), the time-constant of the glass-electrode system, measured from the end of the delay time, is very dependent on the cleanliness of the surface of the glass electrode and on the presence of acid base buffer. Therefore, the electrode was cleaned before the experiment by rubbing with a soft paper tissue soaked with a 0.025% (v/v) solution of Triton X-100 followed by generous rinsing with 120 mM KCI solution and with water. The freshly cleaned glass electrode had a rapid (time constant 0.8 – 0.9 sec) and linear response to several test additions of 125 nmol/ml HCl in a medium containing 3.3 mM glycyl-glycine. The composition of the medium is shown in the Table 13.4. In the absence of glycyl-glycine, the time-constant increased about tenfold (Mitchell & Moyle 1967).

Some technical Commentaries.

Commentary 1. Regular combination pH electrodes that combine pH sensitive glass tip and the Ag/AgCl reference electrode are impractical. This is because the glass tip has high electrical resistance and prone to signal noises if the electrical resistance between the glass and the reference electrodes increases. The porcelain filter of the combination electrode quickly becomes clogged with protein or other impurities, and this greatly enhances the signal noises. It is much cheaper and more reliable to use a mono pH microelectrode. In this case the reference Ag/AgCl commercial electrode is placed in a separate 50 ml centrifugation tube connected with the incubation chamber with the agar-agar-KCL bridge (see in Chapter 11 measurement of membrane potential). The agar-agar-KCl bridge has low electrical resistance and therefore the pH signal is much more stable. In addition, mono-pH electrodes are smaller, much cheaper and more reliable.

Commentary 2. When measuring pH, the electrical signal noise and other troubles are most often caused by problems in electrical connections. All instruments, and magnetic stirrer must be well grounded.

Commentary 3. If you measure simultaneously mitochondrial membrane potential, you must use two separate Ag/AgCl reference electrodes with two separate agar-agar-KCl bridges.

Commentary 4. There was usually observed a slow pH drift in mitochondrial suspensions due to metabolic activity in the mitochondria. This drift is usually- sufficiently slow and constant to permit pH extrapolation over time-intervals of several minutes (see Mitchell & Moyle 1967).

Table 13.5. Troubleshooting of excessive noise and lack of signal with the pH and TPP$^+$-sensitive electrodes.

Cause of the Trouble	Troubleshooting
-A bubble of air, crystals of KCl or broken agar-agar gel in the connection bridge.	-Remove bubbles and fill the tube with KCl, or replace the agar-agar tips.
-KCl crystals formed on the surface of the tube containing Ag/AgCl reference electrode, or other surfaces that might short the electrical circuits.	-Clean all the surfaces of tubes and electrodes
A wire connection is loose or broken.	-Re-solder the connections
-The electrodes are not immersed fully or there is a bubble of air at the electrode's tip.	-Lower the electrodes, -Shake the electrode
-Magnetic stirrer is absent or not rotating.	-Place magnetic rod into the chamber.
- No change in signal	- Check the glass electrode for a crack

Preparation of the agar-agar-KCl connection bridge. Dilute in 5 ml of 3 M KCl 150 mg of agar-agar. Heat in the boiling water bath with the closed cap, and after the solution becomes completely transparent and even, fill the small glass tubes with the gel (see Chapter 8 for more details) and the connecting Tygon tube with 3 M KCl.

Calculations of the H$^+$/Ca^{2+} ratios and Calcium Retention Capacity.

Definition of pH. According to Mitchell & Moyle (1967) the pH of a solution is defined by the equation (1):

$$pH = -\log[H^+] \times f_{H^+} \quad (1)$$

Definition of pH is practically equivalent to that based on H$^+$ ion activity, $[H^+] \times f_{H^+}$. The symbol f_{H^+} represents the abstract H$^+$ ion activity coefficient, which is approximately equal to the mean ionic activity coefficient. Following Mitchell & Moyle (1967), the value of f_{H^+} has been taken as 0.75 in the 150 mM KCl medium, which fits the composition of the medium in our studies.

Computation of buffering capacities in pulsed acid-base titrations. The buffering capacity of the mitochondrial suspensions in glycyl-glycine buffer is almost independent of pH within a given range of 0.1 pH unit over most of the titration range (Mitchell & Moyle 1967). It was therefore generally permissible to equate the virtually instantaneous pH changes of the outer medium (ΔpH_o) caused by injection of a pulse of acid or alkali (within the range of 0.1 pH unit) with the amount of H^+ ion (ΔH^+_o) effectively added to or removed from the phase in rapid equilibrium with the mitochondrial suspension medium (called the outer phase). It was thus possible to describe an outer-phase buffering bower (B_o) defined by equation (2):

$$(\partial H^+/\partial pH)_w + B_o = - \partial(H^+_o)/\partial(pH_o) = - \Delta H^+_o/\Delta pH_o \quad (2)$$

where the first term on the left represents the change of free H^+ ion content of the outer phase with pH, or the capacity of this watery medium (subscript w) to contain free H^+ ions, given by:

$$(\partial H+/\partial pH)w = - 2.303 \, [H+] \times fH+ \quad (3)$$

Eqn. (3) was obtained by differentiating eqn. (1) (Mitchell & Moyle 1967). In the absence of added buffer, B_o represents buffering capacity of the outer phase contributed by the mitochondria only (B^m_o). In the presence of added buffer:

$$B_o = B^m_o + B^b_o \quad (4)$$

where B^b_o is the buffering power of the added buffer. When time was allowed to permit acid-base equilibration between the outer phase and any other phase equilibrating relatively slowly with respect to H^+ and OH^- ions, it was possible to equate a total buffering power, B_T, with the sum of the outer-phase buffering power and inner-phase buffering power, or:

$$B_T = B^m_o + B^b_o + B^m_i \quad (5)$$

Calibrated HCl solutions for pulses and titrations. For calibration and calculations of the Buffer capacities of incubation media it is necessary to use the precisely titrated HCl solutions available commercially (Sigma, Co.). The precise solution of 0.1N HCl is mixed with the equal volume of 200 mM KCl, giving final [HCl] of 0.05N. 10 μl of this 0.05N HCl added to 2 ml of the incubation medium (this volume is sufficient to submerge the tip of the pH microelectrode) was equivalent to addition of 250 nmol/ml H^+ that usually caused in the medium (described in Table 13.4) $\Delta pH = 0.07$.

The Fig. 13.9 illustrates the method of calculation of the H^+/Ca^{2+} ratios and the total amount of calcium sequestered by mitochondria before opening of mPTP. The moment of permeability transition was registered as the beginning of alkalization of the incubation medium.

The calculations are performed using equations presented above. The most important equation is:

$$B_o = \Delta H^+_o/\Delta pH_o \ (6).$$

As an example, let us calculate the H^+/Ca^{2+} ratios for the data presented in Fig. 13.9.

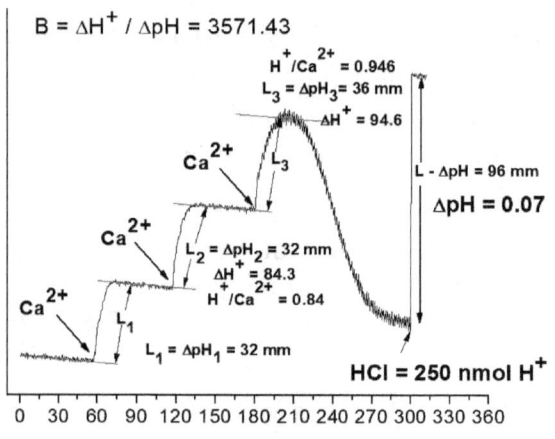

Time (sec.)

Figure 13.9. Graphical method of calculation of the mitochondrial H^+/Ca^{2+} ratios and calcium retention capacities.
The incubation medium contained: KCl 120 mM, NaCl 10 mM, MgCl$_2$ 0.5 mM, KH$_2$PO$_4$ 1 mM, glycyl-glycin 3 mM, pH 7.2, succinate 5 mM, 20 μM ADP + 2 μM oligomycin, rat liver mitochondria (Lewis) 0.5 mg/ml., volume 2 ml. Additions: Ca^{2+} aliquots contained 100 nmol/ml., 5 μl 0.05N HCl per 1 ml contained 250 nmol H^+/ml.

Step 1. First, determine the buffer capacity of the mitochondrial suspension, which is done **after** the mPT occurs, that is we measure the B$_T$. For this, we calibrate the pH trace by adding a known amount of 0.05N HCl. In our examples we added 5 μl of 0.05N to 1 ml, which gives final concentration of 250 μM HCl. For calculations of B, ΔH^+ is given in equivalents (N), and for protons it strictly corresponds to concentration in M, that means that 250 μM correspond to 0.00025 N, or normalized for 1 ml, 250 nmol H^+/ml. Thus B$_T$ = 250 nmol H^+/ΔpH (in our case 0.07) = 250/0.07 = 3571.43.

Step 2. Calculate ΔpH, which occurs after each addition of Ca^{2+} aliquot. For this, we draw the lines as shown in the figure 13.9. As mentioned above, there is often a slight drift of pH caused by the metabolic activity of mitochondria or pH difference between the matrix and the medium. So far, as it is slow and constant it does not interfere with the calculations. Determine the distances in mm between the pH steady states established before and after the Ca^{2+} additions (l_1, l_2, l_3). To calculate ΔpH_1, ΔpH_2, etc. measure the length in mm of the ΔpH after addition of 5 µl calibrated HCl, and calculate the cost of 1 mm in ΔpH: 0.07/96 mm (the lengths were measured from the computer screen), thus 1 mm is equivalent to 0.000729 ΔpH. Next, calculate the ΔH^+ for each addition of Ca^{2+}. Thus, the correspondingly: ΔpH_1 = 32 mm x 0.000729 = 0.0233. From the equation (6) ΔH^+ = B x ΔpH it follows that ΔH^+_1 = 3571.43 x 0.0233 = 83.33 nmol H^+/ml; ΔH^+_2 = 83.33 nmol H^+/ml; ΔH^+_3 = 93.75 nmol H^+/ml.

Step 3. Calculation of H^+/Ca^{2+} ratios for each Ca^{2+} addition: $(H^+/Ca^{2+})_1$ = 83.33 nmol H^+/100 nmol Ca^{2+} = 0.83; $(H^+/Ca^{2+})_2$ = 83.33 nmol H^+/100 nmol Ca^{2+} = 0.83; $(H^+/Ca^{2+})_3$ = 93.75 nmol H^+/100 nmol Ca^{2+} = 0.94.

The corresponding calculations for the data of Fig. 13.7B gave the following values of the H^+/Ca^{2+} ratios: $(H^+/Ca^{2+})_1$ = 0.84; $(H^+/Ca^{2+})_2$ = 0.84; $(H^+/Ca^{2+})_3$ = 0.946.

The examples presented in Figure 13.9 illustrate some events, which must be taken into consideration when calculating the H^+/Ca^{2+} ratios and calcium retention capacities. In Fig. 13.9 the first and second additions of Ca^{2+} gave the same values: $\Delta H^+_{1,2}$ =83.33 nmol H^+/ml, and the $(H^+/Ca^{2+})_{1,2}$ ratio of 0.83. After the third addion of the Ca^{2+} aliquot, $(H^+/Ca^{2+})_3$ ratio = 0.94, which is significantly higher than for the previous two calcium additions. What does it mean? That more cxalcium was sequestrated as quasi-insoluble CaPi salt? Of course not, because soon after the third addition of calcium the large permeability transition pore opened. In this case, the high valkue of ΔH^+_3 indicates that extra number of protons were extruded from protons due to Ca^{2+} cycling that often occurs just before the opening of mPTP. Here, I have to mention that the liver mitochondria were titrated with aliquots containing large amount of calcium. That is, when normalized for 1 mg protein, every addition of brought 200 nmol Ca^{2+}. That is very large amount for the liver mitochondria.

It is clear, that sequestration of CaPi does not occur instantly. During the calcium cycling, Ca^{2+} is transported electrogeniccally, and in the case of a large calcium load on the matrix side of inner membrane the $[Ca^{2+}]_{Free}$ may be close to the critical value, which opens the pore, and unsequestered Ca^{2+} leaves mitochondria. However, because in the protected by ADP + oligomycin mitochondria much of the sequestered CaPi is in the quasi-insoluble form, the matrix concentration of the $[Ca^{2+}]_{Free}$ dropped, and the mPTP closed. As the result, the membrane increased and the extramitochondrial calcium was transported into mitochondria again extruding more protons. Eventually, the cycling of Ca^{2+} sonn resulted in the collapse of the membrane potential and opening of the mPTP. This because liver mitochondria are incapable of maintaining energization due to relatively low rates of respiration. Unprotected liver mitochondria have low capacity for CaPi sequestration in comparison with brain and heart mitochondria.

Addition of ADP and oligomycin protects mitochondria, and thus mitochondria are able to sequester significantly more CaPi (Panov et al. 2004). In the next section we will discuss mechanisms of this protection in more details. Figure 13.9 demonstrates that protected RLM sequester CaPi at higher H^+/Ca^{2+} ratio of 0.84, which indicate that RLM sequester more CaPi as quasi-insoluble $Ca_3(PO4)_2$.

Calcium retention capacity of mitochondria. The calcium retention capacity (CRC) values are calculated by summation of the added Ca^{2+} aliquots and approximation of the amount of Ca^{2+} consumed after the last addition, which caused the permeability transition. For this, you can use the parameters of ΔH^+ from the previous addition of Ca^{2+}.

Figure 13.7 shows experiments with the mouse liver (MLM) (Fig. 137A) and brain mitochondria (Fig. 13.7B) subjected to sequential loads with Ca^{2+} aliquots until the opening of mPTP. Using the pH method, we calculated the H^+/Ca^{2+} ratios for the unprotected mitochondria. The average H^+/Ca^{2+} ratio for the MLM was 0.81 ± 0.01, while for RBM the ratio was 0.93 (Panov et al. 2007). The higher values for H^+/Ca^{2+} ratios of MLM, as compared to RLM, are because in general mouse liver and brain mitochondria have significantly higher rates of respiration with all substrates. The even higher H^+/Ca^{2+} ratio (0.93) obtained with MBM during Ca^{2+} loading, means that most of the calcium phosphate sequestered by MBM was in the quasi-insoluble $Ca_3(PO4)_2$ form; therefore, upon de-energization the concentration of "free" Ca^{2+} in the matrix may be not high enough to cause mPTP to open quickly. Thus, even small changes in the H^+/Ca^{2+} ratios may be responsible for large differences in the total amount of CaPi sequestered by mitochondria from different organs and between the species.

Regulation of the mitochondrial permeability transition.

For several decades regulation of the mitochondrial permeability transition was one of the "hottest" problems in Mitochondriology. Hundreds and hundreds of papers were published on the effects of various conditions and compounds on mitochondrial Ca^{2+} transport and permeability transition. Here I give only a very brief account on this problem referencing my papers (Panov et al. 2004, 2007). In these papers I used the quantitative approach to evaluate and compare the major effectors of mPT. This allowed separating the effects of drugs that influenced mitochondrial energization, from those drugs that influenced the pore-forming protein. In the way, I provided evidence against the possibility that mitochondrial adenine nucleotide translocase is the mPTP.

Normally, mitochondria contain a remarkably constant low concentration of matrix free Ca^{2+} ions (1-5 µM) (Chalmers, Nicholls, 2004). This is important mechanism for regulation of several mitochondrial dehydrogenases by changes in the matrix $[Ca^{2+}]_{Free}$ to comply with the varying energy demands of a tissue (McCormack et al. 1990; Hansford 1985). However, under some pathological conditions accompanied by increased cytosolic calcium concentration, mitochondria may accumulate large amounts of calcium and inorganic phosphate (Pi), which results in mitochondrial dysfunctions followed by apoptotic or necrotic cell death (Halestrap et al. 1998; Crompton 1999).

There is a convincing evidence that functionally mPTP can be considered as a complex consisting of several proteins localized in different mitochondrial compartments: an unidentified, putative pore-forming protein and the adenine nucleotide translocase (ANT) are localized in the inner mitochondrial membrane; porin is located in the outer mitochondrial membrane; and cyclophilin D (CyP-D) is a soluble matrix protein (reviewed in Halestrap et al. 1998; Crompton 1999). ANT and porin form the so-called mitochondrial contact sites between the inner and the outer membranes. Mitochondrial hexokinase and creatine kinase, and pro- and anti-apoptotic members of the Bcl-2 family proteins are also localized at the mitochondrial contact sites. It is more likely than not, that in various tissues the composition and contributions of proteins comprising the above functional complexes controlling the mPT are different. Moreover, the brain tissue does not contain enough calcium to cause opening of mPTP (Panov et al. 2010, 2012). That is why after severe brain trauma the apoptotic death of neurons is postponed, thus giving both patient and a physician time for therapeutical intervention. In comparison, spinal cord tissue contains 8 times more total Ca^{2+} than the brain, and after severe spinal trauma, even without physical injury of the spinal cord, the neurons die by necrosis within the first hours after the trauma (Panov et al. 2012).

Since the ligands of ANT exert powerful control over mPT, it was presumed that ANT forms the large pore of the mPTP (Crompton 1999; Bernardi 1999; Halestrap et al. 2002). This hypothesis was supported by reconstruction experiments with purified ANT, which showed that in the presence of high [Ca^{2+}], ANT can become a non-selective pore (Brustovetsky & Klingenberg 1996). Beutner et al. (1998) showed that in skeletal muscle and brain mitochondria mPTP could be regulated by the structure of creatine kinase located between the inner and outer membranes. However, there is also evidence that mitochondrial proteins other than ANT might form a high conductance pore (Bernardi 1999; Kushnareva et al. 1999).

Among a diverse group of compounds that modify sensitivity of mitochondria to the Ca^{2+} and Pi-induced permeability transition, the best defined are cyclosporin A, and the specific ligands of the adenine nucleotide translocase (ANT). The ANT functions in two conformations: the *c-state* (in which the substrate binding site faces the cytoplasmic side of the inner membrane) and the *m-state* (in which the substrate binding site faces the matrix) (Klingenberg et al. 1973; Panov et al. 1980). Ligands that bind to the m-state of ANT, such as bongkrekic acid and ADP, inhibit the pore, while atractyloside (ATR) and carboxyatractyloside (CATR), which fix ANT in the c-state (Klingenberg et al 1973), are promoters of PT (Panov et al. 1980; Bernardi 1999; Halestrap et al. 2002). These data provide grounds for speculation that only c-state of the ANT is susceptible to Ca^{2+} - induced pore formation. In micromolar concentrations, ADP (in the presence of oligomycin) has an inhibitory effect on the pore. Local anesthetic, dibucain, can enhance the effect of ADP by diminishing the nonspecific proton conductance of the inner mitochondrial membrane (Panov et al. 1980) through stabilizing mitochondrial membranes (Scarpa & Lindsey 1972).

Cyclosporin A is the most potent specific inhibitor of mPTP (Crompton 1999; Bernardi 1999; Halestrap et al. 2002). ADP significantly increases the binding of CsA to mitochondria, and enhances the effect of CsA on mPTP. Remacably, with brain mitochondria CsA is ineffective, if there is no ADP in the external medium. Ca^{2+} decreases the inhibitory action of CsA on the pore by affecting the binding of CsA to the mitochondrial pore "receptor". The intramitochondrial protein receptor was identified in RLM using photoactive CsA derivative and exploring the opposing effects of Ca^{2+} and ADP on pore activity and the photolabeling, thus confirming that CyP-D is the mitochondrial pore receptor for CsA (Crompton 1999). So far, the effects of mPT modifiers have been described qualitatively rather than quantitatively.

Panov et al. (2004, 2007) have shown that opening of the mPTP occurs when mitochondria become less energized and the sequestration process reverses which results in the increase of the matrix $[Ca^{2+}]_{Free}$. There is strong evidence that the interaction of the pore-forming protein(s) with Ca^{2+} and CsA does not require energy (Hansson et al. 2003). Hansson et al (2003) showed that the protective effect of CsA on the Ca^{2+}-induced swelling was dependent on the concentration of $[Ca^{2+}]_{Free}$, and was completely overcomed by $[Ca^{2+}]$ higher than 200 µM. In de-energized rat liver and heart mitochondria, CsA promoted closing of the mPTP, and this effect was enhanced by ADP and Mg^{2+} (Hansson et al. 2003). Hunter and Haworth (1979) found that ADP exerts its effect on the mPTP by interacting with mitochondria at two different sites. One is CATR-sensitive with the high affinity for ADP. It is believed that this site corresponds to the ANT (Hunter & Haworth 1979, Panov et al. 1980). The second site has much lower affinity for ADP and is insensitive to CATR.

In rat liver mitochondria, CATR abolished (i) the specific effects of ADP + oligomycin, (ii) the synergistic effect of ADP added to CsA, and (iii) the synergistic effects of ADP + CsA when added together with dibucain. This is because CATR promotes the c-conformation of the ANT, which enhances conductivity of the ANT channel to protons (Panov et al. 1980). The presence of ADP diminishes H^+ and K^+ conductivity of the mitochondria by conversion of the ANT into m-state (Panov et al. 1980), an effect that tends to enhance mitochondrial energization. The effect of ADP is blocked or reversed by CATR. In the presence of ADP, ANT acquires the m-conformation, which has a lower proton conductance (H^+ conductivity decreases by 50%); CATR converts ANT to the c-conformation and H^+ conductivity increases (Panov et al. 1980). Moreover, there is evidence that only the c-conformation is sensitive to Ca^{2+} induced the pore opening (Halestrap & Davidson 1990).

Thus, the synergistic effect of a low concentration of ADP and CsA may be explained by a combination of (i) increased energization of mitochondria (due to ADP-induced conversion of ANT to the m-conformation state), (ii) the decreased Ca^{2+} sensitivity of the m-conformation of the ANT, and (iii) the lower sensitivity of the mPTP to Ca^{2+} caused by CsA. Additionally, both ADP and CsA may mask the Ca^{2+} binding site of the ANT (Halestrap et al. 2002). Moreover, dibucain diminishes the nonspecific H^+ conductivity of the inner mitochondrial membrane of RLM (H^+ conductivity decreased by 40% at 200 μM dibucain), and the combined effect of dibucain and ADP decreases H^+ conductance by 300% (Panov et al. 1980).

In rat brain mitochondria, the effects of CATR were much more profound. CATR abolished the protection afforded by all mPTP inhibitors (including CsA) and their combinations (Panov et al. 2007). The reasons for this are not fully understood, but may be linked to the amount of ANT isoform levels expressed in the brain in comparison with the liver. It has been shown that total ANT expression (ANT1 plus ANT2) was at least 10-fold higher in the brain than liver (Dorner et al. 1999). Furthermore, ANT1 is the predominant isoform expressed in brain, and ANT2 is the predominant isoform in the liver (Dorner et al. 1999). There is evidence that ANT1 has a higher affinity for CyP-D than ANT2 (Vyssokikh et al. 2001). Panov et al. (2007) suggested that the increased proton conductivity of RBM induced by CATR caused acidification of the matrix and thereby reversed the Ca^{2+} sequestration process. As a result, the concentration of the soluble form of CaPi in the matrix increased sufficiently to abolish the inhibitory effect of CsA on the mPT. Thus, ANT and its specific ligands may control opening of mPTP by affecting energization of mitochondria and the CaPi sequestration process. Panov et al. (2004, 2007) suggested the existence of two different mechanisms of pharmacological regulation of mPT: 1) through influencing the ANT conformational state, which affects mitochondrial energization as with ADP + oligomycin, CATR, bongkrekic acid, local anesthetics; 2) non-energy-dependent effects of

CsA and Ca^{2+} on the mPTP through binding to intramitochondrial CsA receptor CypD (Hansson et al. 2003).

A separation of these two major controlling mechanisms of the mitochondrial permeability transition may allow a better understanding of pathogenic mechanisms of diseases associated with mitochondrial dysfunction, and virtually finished the constant flow of papers on the effect of various modifiers on mPTP.

ATP-synthase complex as a pore-forming structure of the mitochondrial Ca^{2+}-dependent permeability transition.

After it became clear that adenine nucleotide translocase is not a pore-forming protein in the phenomenon of mPT, there was for a significant decline in publications on the subject. But the problem remained open: what protein is responsible for the mitochondrial permeability transition? Recently, there were published several articles, which presented a number of evidences that the mitochondrial ATP-synthase might be the structure responsible for the formation of the mitochondrial permeability transition pore (Bonora et al., 2013; Giorgio et al., 2013). The detailed experimental work of several laboratories suggested that out of all participants of the ATP-synthase complex, the "c" subunit is the only transmembrane structure with the ability to form a high conductance pore. The "c" subunit is capable of oligomerization with formation of the "c"-ring possessing the properties of controlled conductance (McGeoch and Palmer, 1999). This idea was supported and further developed by other authors (2014; Alavian et al., 2014; Azarashvili et al., 2002, 2014; Bonora & Pinton, 2014; Morciano et al., 2014). In experiments with the isolated and purified ATP-synthase complex and "c"-subunit with the subsequent reconstruction in artificial membrane, there were reproduced most of the regulatory properties known for the mPTP (Bonora et al., 2013; Alavian et al., 2014; Morciano et al., 2014), including binding cyclophilin D, cyclosporin A, ADP and the mitochondrial proteins that control apoptosis (Giorgio et al., 2013; Alavian et al., 2014; Morciano et al., 2014).

These new works undoubtedly are of a great importance for understanding of the mPT and the roles mitochondria play in pathogeneses of many diseases. It should be noted that our "old" knowledges on regulation of mPTP by ligands of ANT take a new significance since ATP-synthase, ANT, Pi-transporter and a number of other proteins, such as hexokinase, and creatine kinase may be part of the functional superstructures.

SUPPLEMENTS

Supplement 1. Purification of Ruthenium Red

Ruthenium Red (RR) is widely used in biochemical and histochemical studies as a rather specific inhibitor of Ca^{2+} transport into mitochondria (Broekemeier et al. 1994). However, the commercial RR will not work well or will not work at all without preliminary purification. Ruthenium Red is a pure crystalline polynuclear ionic complex between oxygen-bridged 3- and 4-valent ruthenium atoms and ammonia. It is usually prepared with chloride as the counter ion.

The empirical formula $[Ru_3O_2(NH_3)_{14}]Cl_6$, Mol. Weight 858.5

The structural formula proposed is: $[(NH_3)_5Ru-O-Ru(NH_3)_4-O-Ru(NH_3)_5]^{6+}$

Fresh and old materials are different, even fresh RR from different sources is different. This is because besides pure RR, there may be present its oxidation product Ruthenium Brown (RB), and also a species absorbing at 734, which has violet color – Ruthenium Violet (RV), as well as other derivatives of ruthenium, which are very stable.

The best method to obtain pure RR is crystallization. Pure RR is soluble enough even at 0°C (3.6% in 0.1M NH_3. Below is the method by Luft (1971) without crystallization which separates RV.

0.25 g of crude RR is ground in a small mortar (Coors alumina mortar, size 2) with a few drops of 0.5M NH_3.

Add more 0.5M NH_3 and transfer the suspension with a Pasteur pipette to a 15 ml graduated centrifuge tube, using more of the 0.5M NH_3 to rinse the mortar, making the volume up to 10 ml

Keep the centrifuge tube at 60°C in a water bath for 30 minutes with frequent agitation of the suspension by squirting the solution forth and back with the pipette.

Cool the centrifuge tube with tap water and centrifuge for 5-10 minutes at 1600 g.

Carefully withdraw the supernatant with a pipette and transfer to a small plastic Petri dish. The remaining residue is RV.

Evaporate the supernatant in a (vacuum) dessicator over non-indicating Drierite (anhydrous $CaSO_4$ is available commercially) in an atmosphere of NH_3. This atmosphere is produced by including into dessicator a small quantity (about 10 g) of ammonium carbonate together with a Petri dish of NaOH pellets (about 100 g) to absorb the CO_2.

When the dish is dry (about 48 hrs of evaporation), transfer the residue in to a vial and use as the "purified" RR.

References.

Luft, J.H. (1971) Ruthenium Red and Violet. I. Chemistry, purification, methods of use for electron microscopy and mechanism of action. Anat. Rec. 171, 347-368

Luft, J.H. (1971) Ruthenium Red and Violet. II. Fine structural localization in animal tissues. Anat. Rec. 171, 369-441

Supplement 2. Suggested substrates concentrations.

Chemical/Substrate	Mol. Wt	[Final]
MOPS Na salt	231.25	10 mM
Ascorbic acid*	176.12	0.5 1.0 mM
TMPD (It must be light gray)*	237.2	0.3 mM
Succinic acid	118.1	10 mM
Succinate di-Na salt, 6 H_2O	270.1	5 mM
Malate (Mono Na)	156.1	2 mM
Glutamic acid Mono Na salt	169.1	20 mM
Malate Mono Na salt	156.1	2 mM
L-Glutamic Acid mono-K	185.2	20 mM
		5 mM
Oxaloacetic Acid* Make daily	132.1	10 µM
KH_2PO_4	136.1	2 mM
K_2HPO_4	174.18	
EDTA	292.2	1 mM
EGTA	380.4	1 mM
ADP Na salt	427.2	150 µM
$LaCl_3$*7 H_2O	371.3	10 µM
Carnitine	197.7	0.2 mM
α-Ketoglutarate Na salt*	168.1	10 mM
Pyruvate Na*	110.0	2.5 mM
TPP^+ (Tetraphenyl-phosphonium) Cl	374.86	10 mM
L-Palmitoyl-carnitine Sigma P-1645 10 mg (C2)	436.1	50 µM

PS. Do not use DL-palmitoyl-carnitine since D-isomer inhibits oxidation of L-palmitoyl-carnitine.

Supplement 3. Suggested concentrations of Inhibitors, Uncouplers and Ionophores

Chemical	Mol. Wt	[Final]
Oligomycin A	791.1	2 μg/mg
Rotenone		1 μg/mg
FCCP Store in the dark	254.2	10^{-7}-10^{-8}
CCCP Store in the dark	204.6	10^{-7}-10^{-8}
Valinomycin	1111.4	10^{-7}
Ruthenium Red	786.35	5×10^{-6}
A23187	523.6	1 μM
Cyclosporin A (CsA) Sigma C3662 p.298	1202.6	0.5 μM
CATR Carboxyatractyloside	847.0	5 μM
ATR Atractyloside	803.0	10 μM
Dibucaine	379.9	0.2 mM
Alamethicine		3.5 μg/ml
tert-Butyl hydroperoxide 70% in water. D = 0.937	90.12	0.25 mM
Nigericin, Na salt	747	50-80 ng/ml
Atpenine A5	366.2	1.0 μM
Malonate-DiNa Make fresh daily	166.04	5 mM

Supplement 3. Suggested concentrations of Inhibitors, Uncouplers and Ionophores (Continued).

Chemical	Mol. Wt	[Final]
5-Aminolevulinic acid	*167.59*	0.1 mM
DMSO Dimethylsulfoxide	*78.13* D=1.1004	
Ca Green-2 8K salt (C3730), cell imperm. Keep in darkness	MW 1666 Mol.Probes Kd=550nM	1 μM
Ca Green-5N 6Ksalt, Cell impermeable C3737 Darkness!	MW 1192 Mol. Probes Kd=14μM	1 μM
Malonate acid	*104.06*	5 mM
KCN		5-10 mM

REFERENCES

Abeles, Moshe. Corticonics: Neural Circuits of the Cerebral Cortex. *Cambridge Univ. Press*, 1991 ISBN 0521374766.

Affourtit C., C. L. Quinlan, M. D. Brand. (2012) Measurement of proton leak and electron leak in isolated mitochondria. *Methods Mol. Biol.* **810**, 165-182.

Андреев А. Ю., Кушнарева Ю. Е., Старков А. А. (2005) Митохондриальный метаболизм свободных радикалов кислорода. *Биохимия*, **70**, 246-264.

Alavian K. N., G. Beutner, E. Lazrove, S. Sacchetti, H. A. Park, P. Licznerski, H. Li, P. Nabili, K. Hockensmith, M. Graham, G. A. Porter, Jr., E. A. Jonas. (2014) An uncoupling channel within the c-subunit ring of the F1FO ATP synthase is the mitochondrial permeability transition pore. *Proc. Natl. Acad. Sci. USA.* **111**, 10580-10585.

Angdisen J., Moore V. D., Cline J. M., Payne R. M., Ibdah J. A. (2005) Mitochondrial trifunctional protein defects: molecular basis and novel therapeutic approaches. *Curr. Drug Targets Immune Endocr. Metabol. Disord.* **5**, 27-40.

Archibald J. M. (2009) The Puzzle of Plastid Evolution. *Current Biol.* **19**, R81–R88.

Atkinson D. E. Cellular energy metabolism and its regulation. *Academic Press*, New York. 1977, pp. 85-107.

Atkinson D. E. (1978) The energy charge of the adenylate pool as a regulatory parameter. Interaction with feedback modifiers. *Biochemistry.* **7**, 4030-3034.

Avadhani N. G., Buetow D. E. (1972) Protein synthesis with isolated mitochondrial polysomes. *Biochem. Biophys. Res. Commun.* **46**, 773-778.

Azarashvili T. S., J. Tyynela, I. V. Odinokova, P. A. Grigorjev, M. Baumann, Y. V. Evtodienko, N. E. Saris. (2002) Phosphorylation of a peptide related to subunit c of the F0F1-ATPase/ATP synthase and relationship to permeability transition pore opening in mitochondria. *J. Bioenerg. Biomembr.* **34**, 279-284.

Azarashvili T., Odinokova I., Bakunts A., Ternovsky V., Krestinina O., Tyynelä J., Saris N.-E. L. (2014) Potential role of subunit c of F0F1-ATPase and subunit c of storage body in the mitochondrial permeability transition. Effect of the phosphorylation status of subunit c on pore opening. *Cell Calcium.* **55**, 69–77. doi:10.1016/j.ceca.2013.12.002

Balazs R. (1965a) Control of glutamate metabolism. The effect of pyruvate. *J. Neurochem.* **12**, 63-76.

Balazs R. (1965b) Control of glutamate oxidation in brain and liver mitochondrial systems. *Biochem. J.* **95**, 497-508.

Barja G. (1999) Mitochondrial oxygen radical generation and leak: sites of production in states 4 and 3, organ specificity, and relation to aging and longevity. *J. Bioener. Biomembr.* **31**, 347-366.

Beghi E., G. Logroscino, A. Chio, O. Hardiman, D. Mitchell, R. Swingler, B. J. Traynor. (2006) The epidemiology of ALS and the role of population-based registries. *Biochim. Biophys. Acta.* **1762**, 1150-1157.

Bernardi P., Vassanelli S., Veronese P., Colonna R., Szabo I., Zoratti M. (1992) Modulation of the mitochondrial permeability transition pore. Effects of protons and divalent cations. *J. Biol. Chem.* **267**, 2934-2939.

Bernardi P. (1999) Mitochondrial transport of cations: Channels, exchangers, and permeability transition. *Physiol. Review.* **79**, 1127-1155.

Betarbet R., Sherer, T.B., MacKenzie, G., Garsia-Osuna, M., Panov, A.V., Greenamyre, J. T. (2000) Chronic systemic pesticide exposure reproduces features of Parkinson's disease. *Nature Neuroscience*, **3**, 1301-1306.

Bett, G. C. L., Rasmussen R. L. (2002) 1. Computer Models of Ion Channels. In Cabo, Candido; Rosenbaum, David S., Eds. *Quantitative Cardiac Electrophysiology.* Marcel Dekker. p. 48. ISBN 0-8247-0774-5.

Beutner G., Ruck A., Riede B., Brdiczka D. (1998) Complexes between porin, hexokinase, mitochondrial creatine kinase and adenylate translocator display properties of the permeability transition pore. Implication for regulation of permeability transition by the kinases. *Biochim. Biophys. Acta.* **1368**, 7-18.

Bonora, M., P. Pinton (2014). Shedding light on molecular mechanisms and identity of mPTP. *Mitochondrion.* **21C**: 11.

Boveris A., Chance B. (1973) The mitochondrial generation of hydrogen peroxide. General properties and effect of hyperbaric oxygen. *Biochem. J.* **134**, 707-716.

Bracken, M. B. (2009) Why are so many epidemiology associations inflated or wrong? Does poorly conducted animal research suggest implausible hypotheses? *Ann. Epidemiol.* **19**, 220-224.

Brand M. D., C. Affourtit, T. C. Esteves, K. Green, A. J. Lambert, S. Miwa, J. L. Pakay, N. Parker. (2004) Mitochondrial superoxide production, biological effects, and activation of uncoupling proteins. *Free Rad. Biol. Med.* **37**, 755–767.

Brand M. D. (1990) The proton leak across the mitochondrial inner membrane. *Biochim. Biophys. Acta.* **1018**, 128-133.

Brand M. D. (1990) The contribution of the leak of protons across the mitochondrial inner membrane to standard metabolic rate. *J. Theor. Biol.* **145**, 267-286.

Brand M. D., Brindle K. M., Buckingham J. A., Harper J. A., Rolfe D. F., Stuart J. A. (1999) The significance and mechanism of mitochondrial proton conductance. *Intern. J. Obesity & Related Metab. Disord.* **23**, Suppl. 6, S4-S11.

Brand M. D., J. L. Pakay, A. Ocloo, J. Kokoszka, D. C. Wallace, P. S. Brookes and E. J. Cornwall (2005). The basal proton conductance of mitochondria depends on adenine nucleotide translocase content. Biochem. J. **392**, 353-362.

Brand M. D. (2010) The sites and topology of mitochondrial superoxide production. *Exp. Gerontol.* 45, 466-472.

Brewer G. J. (2007) Iron and copper toxicity in diseases of aging, particularly atherosclerosis and Alzheimer's disease. *Exp. Biol. Med.* **232**, 323-335

Broekemeier K. M., Krebsbch R. J., Pfeiffer D. R. (1994) Inhibition of the mitochondrial Ca^{2+} uniporter by pure and impure ruthenium red. *Mol. Cell. Biochem.* **139**, 33-40

Brown G. C., Brand M. D. (1991) On the nature of the mitochondrial proton leak. *Biochim. Biophys. Acta.* **1059**, 55-62.

Brown G. C (1992) Control of respiration and ATP synthesis in mammalian mitochondria and cells. *Biochem. J.* **264**, 1-13.

Brown G. C. (1992) The leaks and slips of bioenergetic membranes. *FASEB J.* **6**, 2961-2965

Brustovetsky N., Klingenberg M. (1996) Mitochondrial ADP/ATP carrier can be reversibly converted into a large channel by Ca^{2+}. *Biochemistry.* **35**, 8483-8488.

Cadenas S., Brand M. D. (2000) Effects of magnesium and nucleotides on the proton conductance of rat skeletal-muscle mitochondria. *Biochem. J.* **348**, 209-213.

Cadenas E., K. J. A. Davies. (2000) Mitochondrial free radical generation, oxidative stress, and aging. *Free Rad. Biol. Med.* **29**, 222-230.

Campbell N. A., Williamson B., Hyden R. J. (2006) *Biology: Exploring Life.* Boston, Massachussetts, Pearson Prentice Hall. ISBN 0-13-250882.

Carafoli E. (2010) The fateful encounter of mitochondria with calcium: how did it happen? *Biochim. Biophys. Acta.* **1797**, 595-606.

Caroni P., Gazzotti P., Vuillemier P., Simon W., Carafoli E. (1977) Ca^{2+} transport mediated by a synthetic neutral Ca^{2+}-ionophore in biological membranes. *Biochim. Biophys. Acta.* **470**, 437-445.

Chalmers S., Nicholls D. G. (2003) The relationship between free and total calcium concentrations in the matrix of liver and brain mitochondria. *J. Biol. Chem.* **278**, 19062-19070

Chance B. (1977) Electron transfer: Pathways, mechanisms, and controls. **46**, 967-980.

Chance B., Williams G.R. (1955a) Respiratory enzymes in oxidative phosphorylation. I. Kinetics of oxygen utilization. *J. Biol. Chem.* **217**, 383-394.

Chance B., Williams G. R. (1955b) Respiratory enzymes in oxidative phosphorylation. II Difference spectra. *J. Biol. Chem.,* **217**, 395-408.

Chance B., Williams G. R. (1955c) Respiratory enzymes in oxidative phosphorylation. III. The steady-state. *J. Biol. Chem.* **217**, 409-427.

Chance B., Williams G. R. (1956) The respiratory chain and oxidative phosphorylation. *Edvanc. Enzymol.* **17**, 65-134.

Chance B., Sies H., Boveris A. (1979) Hydroperoxide metabolism in mammalian organs. *Physol. Review.* **59**, 527-605.

Chappell J. B., Perry S. V. (1954) Biochemical and osmotic properties of skeletal muscle mitochondria. *Nature.* **173**, 1094-1095.

CRC Handbook of Chemistry and Physics.

Crompton M., Costi A. (1990) A heart mitochondrial Ca2+-dependent pore of possible relevance to reperfusion-induced injury. Evidence that ADP facilitates pore interconversion between the closed and open states. *Biochem. J.* **266**, 33-39.

Crompton M. (1999) The mitochondrial permeability transition pore and its role in cell death. *Biochem. J.* **341**, 233-249

Davey P. J, Haslam J. M., Linnane A. W. (1970) Biogenesis of mitochondria. 12. The effects of aminoglycoside antibiotics on the mitochondrial and cytoplasmic protein-synthesizing systems of Saccharomyces cerevisiae. *Arch. Biochem. Biophys.* **136**, 54-64.

Demel R. A., J. W. Jansen, P. W. van Dijck, L. L. van Deenen. (1977) The preferential interaction of cholesterol with different classes of phospholipids. *Biochim. Biophys. Acta.* **465**, 1-10.

Denton R. M., D. A. Richards, J. G. Chin. (1978) Calcium ions and the regulation of NAD+-linked isocitrate dehydrogenase from the mitochondria of rat heart and other tissues. *Biochem. J.* **176**, 899-906

Desaulniers N., T. S. Moerland, B. D. Sidell. (1996) High lipid content enhances the rate of oxygen diffusion through fish skeletal muscle. *Am. J. Physiol.* **271**, R42-R47.

Dikalov S. I., Vitek M. P., Mason, R. P. (2004) Cupric-amyloid beta peptide complex stimulates oxidation of ascorbate and generation of hydroxyl radical. *Free Radic. Biol. Med.* **36**, 340-347.

Dlaskova A., L. Hlavata, J. Jezek, P. Jezek. (2008) Mitochondrial Complex I superoxide production is attenuated by uncoupling. *Int. J. Biochem. Cell. Biol.* **40**, 2098-2109.

Dorner A., Olesch M., Giessen S., Pauschinger M., Schultheiss H. P. (1999) Transcription of the adenine nucleotide translocase isoforms in various types of tissues in the rat. *Biochim. Biophys. Acta.* **1417**, 16-24.

Dzikovski B. G., V. A. Livshits, D. Marsh. (2003) Oxygen permeation profile in lipid membranes: comparison with transmembrane polarity profile. *Biophys. J.* **85**, 1005-1012.

Ebert D., R. G. Haller, M. E. Walton. (2003) Energy contribution of octanoate to intact rat brain metabolism measured by 13C nuclear magnetic resonance spectroscopy. *J. Neurosci.* **23**, 5928-5935.

Erecinska M., Nelson D., Silver I. A. (1996) Metabolic and energetic properties of isolated nerve ending particles (synaptosomes). *Biochim. Biophys. Acta.* **1275**, 13-34.

Erecinska M., I.A. Silver. (1989) ATP and brain function. *J. Cereb. Blood Flow. Metab.* **9**, 2-19.

Erecinska M., I. A. Silver. (1990) Metabolism and role of glutamate in mammalian brain. *Prog. Neurobiol.* **35**, 245-296.

Erecinska M., I. A. Silver. (1994) Ions and energy in mammalian brain. *Prog. Neurobiol.* **43**, 37-71

Erecinska M., I. A. Silver. (2001) Tissue oxygen tension and brain sensitivity to hypoxia. *Respir. Physiol.* **128**, 263-276.

Erecinska M., M. M. Zaleska, I. Nissim, D. Nelson, F. Dagani and M. Yudkoff. (1988) Glucose and synaptosomal glutamate metabolism: Studies with [15N]glutamate. *J. Neurochem.* **51**, 892-902.

Fontaine E., O. Eriksson, F. Ichas, P. Bernardi. (1998) Regulation of the permeability transition pore in skeletal muscle mitochondria. Modulation By electron flow through the respiratory chain complex I. *J. Biol. Chem.* **273**, 12662-12668.

Fridovich I. (1997) Superoxide anion radical ($O_2^{\cdot-}$), superoxide dismutases, and related matters. *J. Biol. Chem.* **272**, 18515-18517.

Gellerich F. N., Z. Gizatullina, et al. (2009) Extramitochondrial Ca2+ in the nanomolar range regulates glutamate-dependent oxidative phosphorylation on demand. *PLoS ONE.* **4**, e8181.

Giacomello M, Drago I, Pizzo P, Pozzan T. (2007) Mitochondrial Ca^{2+} as a key regulator of cell life and death. *Cell Death Differ.* **14**, 1267-1274.

Giorgio V., S. von Stockum, M. Antoniel, A. Fabbro, F. Fogolari, M. et al. (2013) Dimers of mitochondrial ATP synthase form the permeability transition pore. *Proc. Natl. Acad. Sci. USA.* **110**, 5887-5892.

Gnaiger, E. (Editor) (2007) *Mitochondrial Pathways and Respiratory Control*, 2nd Ed., MiPNet Publications, Innsbruck, Austria.

Gornall, A. G., A. J. Bardawill, M. M. David. (1949) Determination of serum protein by means of the Biuret reaction. J. Biol. Chem. **177**, 751-766.

Greenawalt J. W., Rossi C. S., Lehninger A. L. (1964) Effect of active accumulation of calcium and phosphate ions on the structure of rat liver mitochondria. *J. Cell Biol.* **23**, 21-38.

Hall J. E., I. Vodyanoy, T. M. Balasubramanian, G. R. Marshall. (1984) Alamethicin. A rich model for channel behavior. *Biophys. J.* **45**, 233-247.

Halestrap A. P, Davidson A. M. (1990) Inhibition of Ca^{2+}-induced large-amplitude swelling of liver and heart mitochondria by cyclosporin is probably caused by the inhibitor binding to mitochondrial-matrix peptidyl-prolyl cis-trans isomerase and preventing it interacting with the adenine nucleotide translocase. *Biochem. J.* **268**, 153-160.

Halestrap A. P., Kerr, P. M., Javadov, S., Woodfield, K.-Y. (1998) Elucidating the molecular mechanism of the permeability transition pore and its role in reperfusion injury of the heart. *Biochim. Biophys. Acta.* **1366**, 79-94.

Halestrap A. P., McStay, G. P., Claarke, S. J. (2002) The permeability transition pore complex: another view. *Biochimie.* **84**, 153-166.

Halliwell B. (1996) Vitamin C: antioxidant or pro-oxidant in vivo? *Free Radic. Res.* **25**, 439-454.

Hansford R. G. (1985) Relation between mitochondrial calcium transport and control of energy metabolism. *Rev. Physiol. Biochem. Pharmacol.* **102**, 1-72.

Hansson M. J., Persson, T., Friberg, H., Keep, M. F., Rees, A., Wieloch, T., Elmer, E. (2003) Powerful cyclosporin inhibition of calcium-induced permeability transition in brain mitochondria. *Brain Res.* **960**, 99-111

Hatefi Y. (1985) The mitochondrial electron transport and oxidative phosphorylation system. *Annu. Rev. Biochem.* **54**, 1015-1069

Haworth R. A., Hunter D.R. (1979) The Ca 2+-Induced Membrane Transition in Mitochondria. Nature of the Ca^{2+} Trigger Site. *Arch. Biochem. Biophys.* **195**, 460-467.

Haworth R. A., Hunter D. R. (1980) Allosteric inhibition of the Ca^{2+}-activated hydrophobic channel of the mitochondrial inner membrane by nucleotides. *J. Membr. Biol.* **54**, 231-236.

Held, P. (2012) *An Introduction to Reactive Oxygen Species. Measurement of ROS in Cells.* White Paper, BioTek Instruments, Inc., P.O. Box 998, Highland Park, Winooski, Vermont 05404-0998 USA.

Hirst J., Sucheta, A., Ackrell, B.A.C., Armstrong, F.A. (1996) Electrocatalytic voltammetry of succinate dehydrogenase: Direct quantification of the catalytic properties of Complex electron-transport enzyme. *J. Am. Chem. Soc.* **118**, 5031-5038.

Hoffman D. L., Salter, J. D., Brookes, P. S. (2007) Response of mitochondrial reactive oxygen species generation to steady-state oxygen tension: implications for hypoxic cell signalling. *Am. J. Physiol. Heart Circ Physiol.* **292**, H101-H108.

Hoffman, D. L., P. S. Brookes (2009). Oxygen sensitivity of mitochondrial reactive oxygen species generation depends on metabolic conditions. *J. Biol. Chem.* **284**, 16236-16245.

Huang B., P. Wu, K. M. Popov, R. A. Harris. (2003) Starvation and diabetes reduce the amount of pyruvate dehydrogenase phosphatase in rat heart and kidney. *Diabetes.* **52**, 1371-1376.

Hunter D. R. Haworth R. A. (1979) The Ca^{2+}-induced membrane transition in mitochondria. *Arch. Biochem. Biophys.* **195**, 453-459.

Hunter D. R. Haworth R. A. (1979) The Ca^{2+}-induced membrane transition in mitochondria. III. Transitional Ca2+ release. *Arch. Biochem. Biophys.* **195**, 468-477.

Jensen B. D., Gunter K. K., Gunter T. E. (1986) The efficiencies of the component steps of oxidative phosphorylation. II. Experimental determination of the equivalence of membrane potential and pH gradient in phosphorylation. *Arch. Biochem. Biophys.* **248**, 305-323.

Kamo N., Muratsugu M., Hongoh R., Kobatake Y. (1979) Membrane potential of mitochondria measured with an electrode sensitive to tetraphenyl phosphonium and relationship between proton electrochemical potential and phosphorylation potential in steady state. *J. Membr. Biol.* **49**, 105-121.

Kamo N., Kobatake Y. (1986) Changes of surface and membrane potentials in biomembranes. *Meth. Enzymol.* **125**, 46-58.

Kelley E. E., N. K. Khoo, N. J. Hundley, U. Z. Malik, B. A. Freeman, M. M. Tarpey. (2010) Hydrogen peroxide is the major oxidant product of xanthine oxidase. *Free Radic. Biol. Med.* **48**, 493-498.

Klingenberg M., Schrere B., Stengel-Rutkowski L., Buchholz M., Grebe K. (1973) Experimental Demonstration of the Reorienting (Mobile) Carrier Mechanism Exemplified by the Mitochondrial Adenine Nucleotide Trnaslocation. In: *Mechanisms in Bioenergetics.* G.F. Azzone,L. Ernster,S. Papa, E. Quagliariello, N. Siliprandi, eds., 1973, AP, NY, pp.257-284.

Kokoszka J. E., K. G. Waymire, S. E. Levy, J. E. Sligh, J. Cai, D. P. Jones, G. R. MacGregor, D. C. Wallace. (2004) The ADP/ATP translocator is not essential for the mitochondrial permeability transition pore. *Nature.* **427**, 461-465.

Kono Y., I. Fridovich. (1982) Superoxide radical inhibits catalase. *J. Biol. Chem.* **257**, 5751-5754.

Kopeikina-Tsiboukidou L., G. Deliconstantinos (1983). Functional changes of rat brain mitochondrial enzymes induced by monomeric cholesterol. *Int. J. Biochem.* **15**, 1403-1407.

Kudin A. P., N. Y-B. Bimpong-Butas, S. Velhager, C. E. Elger, W. S. Kunz. (2004) Characterization of superoxide-producing sites in isolated brain mitochondria. *J. Biol. Chem.* **279**, 4127-4135.

Kudin A. P., D. Malinska, W. S. Kunz. (2008) Sites of generation of reactive oxygen species in homogenates of brain tissue determined with the use of respiratory substrates and inhibitors. *Biochim. Biophys. Acta.* **1777**, 689-695

Kushnareva Yu. E., Campo M. L., Kinnally K. W., Sokolove P. M. (1999) Signal presequences increase mitochondrial permeability and open multiple conductance channel. *Arch. Biochem. Biophys.* **366**, 107-115.

Kussmaul L., Hirst, J. (2006) The mechanism of superoxide production by NADH:ubiquinone oxidoreductase (complex I) from bovine heart mitochondria. *Proc. Natl. Acad. Sci. USA*, **103**, 7607-7612.

Labajova A., A. Vojtiskova, P. Krivakova, J. Kofranek, Z. Drahota and J. Houstek (2006). Evaluation of mitochondrial membrane potential using a computerized device with a tetraphenylphosphonium-selective electrode. *Anal. Biochem.* **353**, 37-42.

Lambert A. J., J. A. Buckingham, H. M. Boysen and M. D. Brand. (2008). Diphenyleneiodonium acutely inhibits reactive oxygen species production by mitochondrial complex I during reverse, but not forward electron transport. <u>*Biochim. Biophys. Acta.*</u> **1777**, 397-403

LaNoue K. F., Strzelecki T., Strzelecka D., Koch C. (1986) Regulation of the uncoupling protein in brown adipose tissue. *J. Biol. Chem.* **261**, 298-305.

LaNoue K. F., Bryla J., Williamson J. R. (1972) Feedback interactions in the control of citric acid cycle activity in rat heart mitochondria. *J. Biol. Chem.* **247**, 667-679.

LaNoue K. F.; Williamson-J. R. (1971) Interrelationships between malate-aspartate shuttle and citric acid cycle in rat heart mitochondria. *Metabolism.* **20**, 119-140.

LaNoue K. F., Bryla J., Bassett D. J. P. (1974) Energy-driven aspartate efflux from heart and liver mitochondria. *J. Biol. Chem.* **249**, 7514-7521.

LaNoue K. F., Jeffries F. M. H., Radda G. K. (1986) Kinetic control of mitochondrial ATP synthesis. *Biochemistry.* **25**, 7667-7675.

LaNoue K. F., Schoolwerth A. C. (1979) Metabolite transport in mitochondria. *Ann. Rev. Biochem.* **48**, 871-922.

LaNoue K. F., Tischler M. E. (1974) Electrogenic characteristics of the mitochondrial glutamate aspartate antiporter. *J. Biol. Chem.* **249**, 7522-7528.

LaNoue K. F., Duczynski J.,Watts J. A., McKee E. (1979) Kinetic properties aspartate transport in rat heart mitochondrial inner membrane. *Arch. Biochem. Biophys.* **195**, 578-590.

LaNoue K. F.,Walajtys E. I., Williamson J. R. (1973) Regulation of glutamate metabolism and interactions with the citric acid cycle in rat heart mitochondria. *J. Biol. Chem.* **248**, 7171-7183.

Lardy H. A., H. Wellman. (1952) Oxidative phosphorylations; role of inorganic phosphate and acceptor systems in control of metabolic rates. *J. Biol. Chem.* **195**, 215-224.

Lee C-P., Ernster L. (1967) Energy coupling in nonphosphorylating submitochondrial particles. In: *Methods in Enzymology*, Estabrook, R.W. and Pullman M.E., eds., pp. 543-548, Academic Press, New York – London.

Lepe-Zuniga J.L., Zigler J.S., Gery I. (1987) Toxicity of light-exposed Hepes media. *J. Immunol. Methods.* **103**, 145. doi:10.1016/0022-1759(87)90253-5. PMID 3655381.

Liochev S. L. (1996) The role of iron-sulfur clusters in in vivo hydroxyl radical production. *Free Radic. Res.* 25 (5), 369-384.

Liochev S. L. (1996) The role of iron-sulfur clusters in in vivo hydroxyl radical production. *Free Radic. Res.* **25**, 369-384.

Lipinski B. (2011) Hydroxyl radical and its scavengers in health and disease. *Oxid. Med. Cell. Longev.* 809696.

Loschen G., Flohe L., Chance B. (1971) *FEBS Letters.* **18**, 261-264.

Loschen G., A. Azzi, L. Flohe (1973) Mitochondrial H_2O_2 formation: Relationship with energy conservation. *FEBS Lett.* **33**, 84-87.

Lou P-H., Hansen B.S., Olsen P.H., Tullin S., Murphy M.P., Brand M.D. (2007) Mitochondrial uncouplers with an extraordinary dynamic range. *Biochem. J.* **407**, 129–140.

Luft, J.H. (1971) Ruthenium Red and Violet. I. Chemistry, purification, methods of use for electron microscopy and mechanism of action. *Anat. Rec.* **171**, 347-368.

Luft, J.H. (1971) Ruthenium Red and Violet. II. Fine structural localization in animal tissues. *Anat. Rec.* **171**, 369-441.

Madden T. D., C. Vigo, K. R. Bruckdorfer, D. Chapman. (1980) The incorporation of cholesterol into inner mitochondrial membranes and its effect on lipid phase transition. *Biochim. Biophys. Acta.* **599**, 528-537.

Marchetti P., Castedo M., Susin S.A., Zamzami N., Hirsh T., et al., (1996) Mitochondrial permeability transition is a central coordinating event of apoptosis. *J. Exp. Med.* **184**, 1155-1160.

Marcus R. A., Sutin N. (1985) Electron transfers in chemistry and biology. *Biochim. Biophys. Acta.* **811**, 265-322.

Massey V. (1994) Activation of Molecular Oxygen by Flavins and Flavoproteins. *J. Biol. Chem.* **269**, 22459-22462.

McGuigan J. A. S., Lutji D., Buri A. (1991) Calcium buffer solutions and how to make them: A do it yourself guide. *Canad. J. Physiol. Pharmacol.* **69**, 1733-1749.

McCormack J. C., Halestrap A. P., Denton R. M. (1990) Role of calcium ions in regulation of mammalian intramitochondrial metabolism. *Physiol.Rev.* **70**, 391-425.

McDonald J.M., Bruns D.E., Jarett L. (1976) Ability of insulin to increase calcium binding by adipocyte plasma membranes (atomic absorption/magnesium/equilibrium analysis). *Proc. Natl. Acad. Sci. USA.* **73**, 1542-1546.

McFadden G. I; Van Dooren G. G (2004) Evolution: Red Algal Genome Affirms a Common Origin of All Plastids. *Current Biol.* **14**, R514–R516.

McGeoch J. E., Palmer D. N. (1999) Ion pores made of mitochondrial ATP synthase subunit c in the neuronal plasma membrane and Batten disease. *Mol. Genet. Metab.* **66**, 387–392.

Michiels C. (2004) Physiological and pathological responses to hypoxia. *Am. J. Pathol.* **164**, 1875-1882.

Mitchell P., Moyle J. (1967) Acid-base titration of rat liver mitochondria. *Biochem. J.* **104**, 588-600.

Missirlis F., Hu J., Kirby K., Hilliker A.J., Rouault T.A., Phillips J.P. (2003) Compartment-specific protection of iron-sulfur proteins by superoxide dismutase. *J Biol Chem.* **278**, 47365-47369.

Miwa S., M. D. Brand. (2003) Mitochondrial matrix reactive oxygen species production is very sensitive to mild uncoupling. *Biochem. Soc. Trans.* **31**, 1300-1301.

Miyadera H., K. Shiomi, H. Ui, Y. Yamaguchi, R. Masuma, et al. (2003) Atpenins, potent and specific inhibitors of mitochondrial complex II (succinate-ubiquinone oxidoreductase). *Proc. Natl. Acad. Sci. USA.* **100**, 473-477.

Moller M., H. Botti, C. Batthyany, H. Rubbo, R. Radi and A. Denicola (2005) Direct measurement of nitric oxide and oxygen partitioning into liposomes and low density lipoprotein. *J. Biol. Chem.* **280**, 8850-8854.

Morciano, G., C. Giorgi, M. Bonora, S. Punzetti, R. Pavasini, et al. (2015) Molecular identity of the mitochondrial permeability transition pore and its role in ischemia-reperfusion injury. J. Mol. Cell. Cardiol. **78**, 142-153.

Moreadith, R., Fiskum, D.R. (1984) Isolation of mitochondria from ascites tumor cells permeabilized with digitonin. Anal. Biochem. **137**, 360-367.

Moser C. C., Page C. C., Farid R., Dutton P. L. (1995) Biological electron transfer. *J. Bioenerg. Biomembr.* **27**, 263-274.

Moser C. C., T. A. Farid, et al. (2006). Electron tunneling chains of mitochondria. *Biochim. Biophys. Acta.* **1757,** 1096-109.

Muller F. L., Liu Y., Van Remmen H. (2004) Complex III releases superoxide to both sides of the inner mitochondrial membrane. *J. Biol. Chem.* **279,** 49064-49073.

Murphy M. P. (2006) Induction of mitochondrial ros production by electrophilic lipids: A new pathway of redox signaling? *Am. J. Physiol. Heart Circ. Physiol.* **290,** H1754-1755.

Murphy M. P., R. A. Smith. (2007) Targeting antioxidants to mitochondria by conjugation to lipophilic cations. *Annu. Rev. Pharmacol. Toxicol.* **47,** 629-656.

Murphy M. P. (2009) How mitochondria produce reactive oxygen species. *Biochem. J.* **417,** 1-13.

Nalecz K. A., Miecz D., Berezowski V., Cechelli R. (2004) Carnitine: Transport and physiological functions in brain. *Molec. Aspects Medicine.* **25,** 551-567.

Nowell, N., Nalty, M.S. (1986) A digitonin-based procedure for the isolation of mitochondrial DNA from mammalian cells. Plasmid. **16,** 77-80.

Ohnishi S. T., Ohnishi O., Muranaka S., Fujita H. Kimura, et al. (2005) A Possible Site of Superoxide Generation in the Complex I Segment of Rat Heart Mitochondria. *J. Bioenerg. Biomembr.* **37,** 1-15.

Pacher P., J. S. Beckman, L. Liaudet. (2007) Nitric oxide and peroxynitrite in health and disease. *Physiol. Rev.* **87,** 315-424.

Panov A. V., Konstantinov Yu. M., Lyakhovich V. V. (1975) The possible role of palmitoyl-CoA in the regulation of the adenine nucleotide transport in mitochondria under different metabolic states. *J. Bioenergetics.* **7,** 75-85.

Panov A., Filippova S., Lyakhovich V. (1980) Adenine nucleotide translocase as a site of regulation by ADP of the rat liver mitochondria Permeability to H^+ and K^+ ions. *Arch. Biochem. Biophys.* **199,** 420-426.

Panov A. V., Solov'ev V. N., Vavilin V. A. (1991) Inter strain differences in organization of metabolic processes in the rat liver. I. The dynamics of changes in the contents of adenine nucleotides, glycogen and fatty acyl-CAs in the course of short-term starvation in the livers of rats of Wistar, August and Wag strains. Intern. *J. Biochem.* **23,** 875-879.

Panov A. V., Scaduto R.C., Jr. (1996) Substrate specific effects of calcium on metabolism of rat heart mitochondria. *Am. J. Physiol.* (*Heart Circ. Physiol. 39*). **270**, H1398-H1406.

Panov A., Scarpa A. (1966a) Independent modulation of the activity of α-ketoglutarate dehydrogenase complex by Ca^{2+} and Mg^{2+}. *Biochemistry* (US), **35**, 427-432.

Panov A., Scarpa A. (1966b) Mg^{2+} control of respiration in isolated rat liver mitochondria. *Biochemistry* (USA). **35**, 12849-12856.

Panov A., Andreeva L., and Greenamyre J.T. (2004) Quantitative evaluation of the effects of mitochondrial permeability transition pore modifiers on accumulation of calcium phosphate: Two modes of action of mPTP modifiers. *Arch. Biochem. Biophys.* **424**, 44-52.

Panov A., S. Lund J. T. Greenamyre. (2005) Ca^{2+}-induced permeability transition in human Lymphoblastoid cell mitochondria from normal and Huntington's disease individuals. *Mol. Cell. Biochem.* **269**, 143-152.

Panov A., Dikalov S., Shalbueva N., Taylor G., Sherer T., and Greenamyre J.T. (2005a) Rotenone model of Parkinson's disease: Multiple brain mitochondria dysfunctions after short-term systemic rotenone intoxication. *.J. Biol. Chem.* **280**, 42026-42035.

Panov A., Dikalov S., Shalbuyeva N., Hemendinger R., Greenamyre J.T., Rosenfeld J. (2007) Species and tissue specific relationships between mitochondrial permeability transition and generation of ROS in brain and liver mitochondria of rats and mice. *Am. J. Physiol. Cell. Physiol.* **292**, C708-C718.

Panov, A., P. Schonfeld, S. Dikalov, R. Hemendinger, H. L. Bonkovsky and B. R. Brooks. (2009) The Neuromediator Glutamate, through Specific Substrate Interactions, Enhances Mitochondrial ATP Production and Reactive Oxygen Species Generation in Nonsynaptic Brain Mitochondria. *J. Biol. Chem.* **284**, 14448-14456.

Panov A. V., V. A. Vavilin, V. V. Lyalkhovich, B. R. Brooks, H.L. Bonkovsky. (2010a) Effects of defatted bovine serum albumin on respiratory activities of brain and liver mitochondria from C57Bl/6G mice and Sprague Dawley rats. *Bull. Exp. Biol. Med.* **149**, 187-190.

Panov A., N. Kubalik, B. R. Brooks, C. A. Shaw. (2010в) In vitro effects of cholesterol β-D-glucoside,cholesterol and cycad phytosterol glucosides on respiration and ROS generation in brain mitochondria. *J. Membrane Biol.* **237**, 71-77.

Panov, A., Kubalik, N., Zinchenko, N., Ridings, D.M., Radoff, D.A., Hemendinger, R., Brooks, B.R., Bonkovsky, H.L. (2011a) Metabolic and functional differences between brain and spinal cord mitochondria underlie different predisposition for pathology. *Am. J. Physiology. Regul. Integr. Comp. Physiol.* **300**, R844-R854.

Panov A., Kubalik N., Brooks B. R. (2010) Effects of palmitoyl carnitine on brain, spinal cord and heart mitochondria from wild type and transgenic SOD1 rats. 21st International Symposium on ALS/MND, Orlando 11-13 December, 2010. *Amyotrophic Lateral Sclerosis* 11 (Suppl. 1), **115.**

Panov A., Dikalov S., Dambinova S. (2011) Tissue-specific metabolic regulations of respiration and ROS production of the heart, brain and spinal cord mitochondria. *FASEB Meeting*, Washington DC, April 10-14, 2011.

Panov A, Steuerwald N., Vavilin V., Dambinova S., Bonkovsky H. L. (2012) Role of neuronal mitochondrial metabolic phenotype in pathogenesis of ALS. In. *Amyotrophic Lateral Sklerosis.* Ed. M.H. Maurer. Pages 225-248, Intech Open Access Publisher. 2012. ISBN 979-953-307-199-1.

Panov A., Z. Orynbayeva, V, Vavilin, and V. Lyakhovich (2014) Fatty Acids in Energy Metabolism of the Central Nervous System. Review Article. BioMed Res. Intern. **Vol. 2014**, Article ID 472459, 22 pages. http://dx.doi.org/10.1155/ 2014/472459.

Panov A., Dikalov S. I. (2014) Chapter 13. Structural and metabolic determinants of mitochondrial ROS and methods of ROS detection. In: *Systems Biology of Free Radicals and Antioxidants.* Laher, L. (Ed.) pp. 296-322. Springer-Verlag Berlin Heidelberg 2014. DOI 10.1007/978-3-642-30018-9_6.

Perry S. W., J. P. Norman, J. Barbieri, E. B. Brown and H. A. Gelbard. (2011) Mitochondrial membrane potential probes and the proton gradient: a practical usage guide. *Biotechniques.* **50**, 98-115.

Perevoshchikova I. V., C. L. Quinlan, A. L. Orr, A. A. Gerencser, M. D. Brand. (2013) Sites of superoxide and hydrogen peroxide production during fatty acid oxidation in rat skeletal muscle mitochondria. *Free Radic. Biol. Med.* **61**, 298-309.

Petit P. X., Susin S-A., Zamzami N., Mignotte B., Kroemer G. (1996) Mitochondria and programmed cell death: back to the future. *FEBS Letters.* **396**, 7-13.

Pivovarova N. B., Andrews S. B. (2010) Calcium-dependent mitochondrial function and dysfunction in neurons. *FEBS J.* **277**, 3622-3636.

Power G.., Stegall H. (1970) Solubility of gases in human red blood cell ghosts. *J. Appl. Physiol.* **29**, 145-149.

Priyadarshi A, Khuder SA, Schaub EA, Priyadarshi SS (2001) Environmental risk factors and Parkinson's disease: a metaanalysis. *Environ Res.* **86**, 122-127.

Quinlan C. L., Orr A .L., Perevoshchikova I. V., Treberg J. R., Ackrell B. A., Brand M. D. (2012) Mitochondrial Complex II Can Generate Reactive Oxygen Species at High Rates in Both the Forward and Reverse Reactions. *J. Biol. Chem.* **287**, 27255–27264.

Quinlan C. L., I. V. Perevoshchikova, R. L. Goncalves, M. Hey-Mogensen, M. D. Brand (2013) The determination and analysis of site-specific rates of mitochondrial reactive oxygen species production. *Methods Enzymol.* **526**, 189-217.

Rau A. R. (2002) Biological scaling and physics. *J. Biosci.* **27**, 475-478.

Reynafarje B., Costa L. E., Lehninger A. L. (1985) Oxygen solubility in aqueous media determined by a kinetic method. *Anal. Biochem.* **145**, 406-418.

Rottenberg H. (1979) The measurement of membrane potential and ΔpH in cells, organelles, and vesicles. *Methods Enzymol.* Vol. LV, 547-569.

Rottenberg H. (1984) Membrane potential and surface potential in mitochondria: Uptake and binding of lipophilic cations. *J. Membr. Biol.* **81**, 127-138.

Rustin P., Chretien, D., Bourgeron, T., Gerard, A., Rotig, A., et al. (1994) Biochemical and molecular investigations in respiatory chain deficiencies. *Clin. Chim Acta.* **228**, 35-51.

Sawyer D. T., Valentine J. S. (1981) How super is superoxide? *Acc. Chem. Res.* **14**, 393-400.

Scarpa A., Lindsey J.G. (1972) Maintenance of energy-linked functions in rat liver mitochondria aged in the presence of nupercaine. *Eur. J.Biochem.* **27**, 401-407.

Schagger H. (2001) Respiratory chain supercomplexes. *IUBMB Life.* **52**, 119-128.

Schagger H., R. de Coo, M.F. Bauer, S. Hoffman, C. Godino, U. Brandt. (2004) Significance of Respirasomes for the

Assembly/Stability of Human Respiratory Chain Complex I. *J. Biol. Chem.* **279**, 36349-36353.

Schagger H., K. Pfeiffer. (2000) Supercomplexes in the respiratory chains of yeast and mammalian mitochondria. *EMBO J.* **19**, 1777-1783.

Schagger H., K. Pfeiffer. (2001) The Ratio of Oxidative Phosphorylation Complexes I–V in Bovine Heart Mitochondria and the Composition of Respiratory Chain Supercomplexes. *J. Biol. Chem.* **276**, 37861-37867.

Shiva S, Brookes PS, Patel RP, Anderson, P.G., Darley-Usmar, V.M. (2001) Nitric oxide partitioning into mitochondrial membranes and the control of respiration at cytochrome c oxidase. Proc. Natl. Acad. Sci. USA. 2001, 98, 7212-7217

Shiva S, Crawford JH, Ramachandran A et al. (2004) Mechanisms of the interaction of nitroxyl with mitochondria. *Biochem J.* **379**, 359-366.

Schonfeld, P., G. Reiser. (2013). Why does brain metabolism not favor burning of fatty acids to provide energy? Reflections on disadvantages of the use of free fatty acids as fuel for brain. *J. Cereb. Blood Flow Metab.* **33**, 1493-1499.

Simon, W., D. Ammann, M. Oehme, W. E. Morf. (1878) Calcium-selective electrodes. *Ann. N.Y. Acad. Sci.* **307**, 52-69.

Simon, W., E. Carafoli. (1979) Design, properties and applications of neutral ionophores. *Methods in Enzymology.* **56**, 439-448.

Sims N. R. (1990) Rapid isolation of metabolically active mitochondria from rat brain and subregions using Percoll density gradient centrifugation. *J. Neurochem.* **55**, 698-707.

Sipos I., L. Tretter, V. Adam-Vizi. (2003) The Production of Reactive Oxygen Species in Intact Isolated Nerve Terminals Is Independent of the Mitochondrial Membrane Potential. Neurochem. Res. **28**, 1575–1581.

Sipos I., Tretter L., Adam-Vizi V. (2003) Quantitative relationship between inhibition of respiratory complexes and formation of reactive oxygen species in isolated nerve terminals. *J. Neurochem.* **84**, 112-118.

Skulachev V. P. (1996) Role of uncoupled and non-coupled oxidations in maintenance of safely low levels of oxygen and its one-electron reductants. *Rev. Biophys.* **29**, 169-202.

Sorgato M.C., Sartorelli L., Loschen G., Azzi, A. (1974) Oxygen radicals and hydrogen peroxide in rat brain mitochondria. *FEBS Letters.* **45**, 92-95.

Stanley C. J., Perham R. N. (1980) Purification of 2-oxo acid dehydrogenase multyenzyme complexes from ox heart by a new method. *Biochemical J.* **191**, 147-154.

Starkov A. A., Fiskum G. (2001) Myxothiazol induces h(2)o(2) production from mitochondrial respiratory chain. *Biochem. Biophys. Res. Commun.* **281**, 645-650.

Starkov A. A., Fiskum G. (2003) Regulation of brain mitochondrial H2O2 production by membrane potential and NAD(P)H redox state. *J. Neurochem.* **86**, 1101-1107.

Starkov A. A. (2008). The role of mitochondria in reactive oxygen species metabolism and signaling. *Ann. NY. Acad. Sci.* **1147**, 37-52.

Starikovskaya E. B., R. I. Sukernik, O. A. Derbeneva, N. V. Volodko, E. Ruiz-Pesini, et al. (2005) Mitochondrial DNA diversity in indigenous populations of the southern extent of Siberia, and the origins of Native American haplogroups. *Ann. Hum. Genet.* **69**, 67-89.

St-Pierre J., Buckingham J. A., Roebuck S. J., Brand M. D. (2002) Topology of superoxide production from different sites in the mitochondrial electron transport chain. *J. Biol. Chem.* **277**, 44784-44790.

Subczynski W. K., J. S. Hyde. (1983) Concentration of oxygen in lipid bilayers using a spin-label method. *Biophys. J.* **41**, 283-286.

Szabo C. (2003) Multiple pathways of peroxynitrite cytotoxicity. *Toxicol. Lett.* **140/141**, 105-112.

Szabo C., Ischiropoulos H., Radi R. (2007) Peroxynitrite: biochemistry, pathophysiology and development of therapeutics. *Nat. Rev. Drug. Discov.* **6**, 662-680.

Szatrowski T. P., C. F. Nathan. (1991) Production of large amounts of hydrogen peroxide by human tumor cells. *Cancer Res.* **51**, 794-798.

Tarpey M. M., I. Fridovich. (2001) Methods of detection of vascular reactive species: nitric oxide, superoxide, hydrogen peroxide, and peroxynitrite. *Circ. Res.* **89**, 224-236.

Tillakaratne N. J., L. Medina-Kauwe, K. M. Gibson. (1995) Gamma-aminobutyric acid (GABA) metabolism in mammalian neural and nonneural tissues. *Comp. Biochem. Physiol. A Physiol.* **112** , 247-263.

Tsien R. Y., T. J. Rink. (1980) Neutral arrier ion-selective microelectrodes for measurement of intracellular free calcium. *Biochim. Biophys. Acta.* **599**, 623-638.

Tretter L., V. Adam-Vizi. (2000) Inhibition of Krebs Cycle Enzymes by Hydrogen Peroxide: A Key Role of a-Ketoglutarate Dehydrogenase in Limiting NADH Production under Oxidative Stress. *J. Neurosci.* **20**, 8972–8979.

Tretter L., Adam-Vizi V. (2004) Generation of reactive oxygen species in the reaction catalyzed by alpha-ketoglutarate dehydrogenase. *J. Neurosci.* 24, 7771-7778.

Tretter L., D. Mayer-Takacs and V. Adam-Vizi (2007a). The effect of bovine serum albumin on the membrane potential and reactive oxygen species generation in succinate-supported isolated brain mitochondria. *Neurochem. Int.* **50**, 139-147.

Tretter L. and V. Adam-Vizi (2007b). Moderate dependence of ROS formation on $\Delta\Psi$ in isolated brain mitochondria supported by NADH-linked substrates. *Neurochem. Res.* **32**, 569-575.

Tretter L., K. Takacs, V. Hegedus and V. Adam-Vizi (2007c). Characteristics of alpha-glycerophosphate-evoked H2O2 generation in brain mitochondria. *J. Neurochem.* **100**, 650-663.

Trnka J., F. H. Blaikie, A. Logan, R. A. J. Smith, and M. P. Murphy. (2009) Antioxidant properties of MitoTEMPOL and its hydroxylamine. *Free Radic. Res.* **43**, 4–12.

Trounce, I.A., Kim, Y.L., Jun, A.S., Wallace, D.C. (1996) Assessment of mitochondrial oxidative phosphorylation in patient muscle biopsies, lymphoblasts, and transmitochondrial cell lines. Methods Enzymol. **264**, 484-509.

Turrens J .F. (2003) Mitochondrial formation of reactive oxygen species. *J. Physiol.* **552**, 335-344.

Urry D. W. (1978) Basic aspects of calcium chemistry and membrane interaction: on the messenger role of calcium. *Ann. NY Acad. Sci.* **307**, 3-27.

Vanacore N, Nappo A, Gentile M et al. (2002) Evaluation of risk of Parkinson's disease in a cohort of licensed pesticide users. *Neurol Sci.* **23**, Suppl 2, S119-120.

Vinogradov A. D., D. Winter, T. E. King. (1972) The binding site for oxaloacetate on succinate dehydrogenase. *Biochem. Biophys. Res. Commun.* **49**, 441-444.

Votyakova T.V., Reynolds I.J. (2001) ΔΨ-dependent and independent production of reactive oxygen species by rat brain mitochondria. *J. Neurochem.* **79**, 266-277.

Vyssokikh M. Yu., Katz A., Rueck A., Wuensch C., Dorner A., Zorov D.B., Brdiczka D. (2001) Adenine nucleotide translocator isoforms 1 and 2 are differently distributed in the mitochondrial inner membrane and have distinct affinities to cyclophilin D. *Biochem. J.* **358**, 349-358.

Weisiger R. A., I. Fridovich. (1973) Mitochondrial superoxide simutase. Site of synthesis and intramitochondrial localization. *J. Biol. Chem.* **248**, 4793-4796.

Windrem D. A., Plachy W. Z. (1980) The diffusion solubility of oxygen in lipid bilayers. *Biochim. Biophys. Acta.* **600**, 655-665.

Wong-Riley M. T. (1989).. Cytochrome oxidase: An endogenous metabolic marker for neuronal activity. *Trends Neurosci.* **12**, 94-101.

Wirmer J., Westhof E. (2006) Molecular contacts between antibiotics and the 30S ribosomal particle. *Methods Enzymol.* **415**, 180-202.

Wang Z., T. P. O'Connor, S. Heshka, S. B. Heymsfield. (2001) The reconstruction of Kleiber's law at the organ-tissue level. *J. Nutr.* **131**, 2967-2970.

Wallace D. C. (1992) Diseases of the mitochondrial DNA. *Ann. Rev. Biochem.* 61, 1175-1212.

Wallace D. C., Shoffner J. M., Trounce I., Brown M. D., Ballinger S., et al. (1995) Mitochondrial DNA mutations in human degenerative diseases and aging. *Biochim. Biophys. Acta.* **1271**, 141-151.

Wallace D. C. (1994) Mitochondrial DNA sequence variation in human evolution and disease. *Proc. Natl. Acad. Sci. USA.* **91**, 8739-8746.

Wallace D. C., Torroni A. (1992) American Indian prehystory as written in the mitochondrial DNA. *Human Biology.* **64**, 403-416.

CONTENTS

Subject   **Page**

simultaneous measurements of respiration and membrane potential with a TPP⁺-sensitive electrode.

Capacity.

SUPPLEMENTS

REFERENCES   220

CONTENTS   239

www.ingramcontent.com/pod-product-compliance
Lightning Source LLC
Chambersburg PA
CBHW051450170526
45166CB00001B/187